Missing Piece in the Puzzle: Unveiling Pedestrian Synchronization on Slender Canopy Walkways

Shiva

Table of contents

Chapter 1

1 Introduction

1.1 Background and problem statement

Modern footbridges are designed with philosophies aimed at improving the aesthetics of the structure while maintaining a cost-effective design (Ingolfsson, et al., 2012). The advanced and sustainable construction materials used in modern footbridges are made to resist higher stresses induced by static loads but this has resulted in slender, and long spanned structures with smaller cross-sectional dimensions   (Ingolfsson, et al., 2012). The consequence of such designs and constructions has lead to the adverse dynamic behaviour of footbridges under pedestrian loading (Brownjohn, Zivanovic & Pavic, 2008).

Increasingly, the concept of slender low mass footbridges has been used in a class of footbridges located in parks and public spaces (Nhleko, 2016). Formally, these structures are known as canopy walkways and serve as an important part of the ecotourism infrastructure by providing pedestrians with access to the forest canopy in several tourist attractions around the world. Because of the need to reduce the visual impact on the surrounding environment, the design and construction of canopy walkways necessitates the use of novel and slender structural systems which result in low natural frequencies and reduced masses (Nhleko, 2016). This makes them susceptible to pedestrian groups or crowd induced vibrations (Newland, 2004 & Ingolfsson, et al., 2012).

Researchers have conducted extensive research on the dynamic behaviour of typical footbridges subjected to pedestrian loading (Ingólfsson, Georgakis & Jönsson, 2012). Additionally, much research effort has been devoted to group or crowd induced vibrations which lead to the phenomenon of synchronous lateral excitation (SLE). This has been observed in structures such as the London Millennium bridge which experienced excessive lateral vibrations on opening day (Dallard, Flint, et al., 2001). Investigations aimed at understanding SLE lead to the development of a model often reffered to as the Arup model, capable of evalutating the critical number of pedestrians required to trigger lateral synchronization and lock-in on a footbridge. However, there is currently a lack of research investigating the evidence of pedestrian lateral synchronization and the lock-in phenomenon on canopy walkways. Furthermore, there are a few, if any, studies devoted to evaluating the critical number of pedestrians required to trigger lateral synchronization or lock-in on canopy walkways.

The significance of this study is in the fact that, to the author's knowledge, there are no reported studies of lateral synchronization and lock-in on canopy walkways. Canopy walkways usually have complex geometries and do not adhere to conventional vibration comfort limits as established by the design guidelines. This is due to client specifications, the need to design light-weight and unobtrusive structures and the need to enhance the experience of the user. These design considerations have considerable implications on the dynamic properties and the dynamic behaviour of the structure under pedestrian loading.

The aim of this book is to conduct crowd related investigations on the "Boomslang" Canopy walkway to evaluate the critical number of pedestrians required to trigger lateral synchronization on the viewing structure. The Boomslang is located in the SANBI Kirstenbosch Gardens, in the southern suburbs of Cape Town. The Arup model used to evluate this result is governed by the assumption that SLE is the initiating mechanism for excessive lateral vibrations on a footbridge. Depending on the results of this study, the author also aims to either support or challenge (with evidence) the notion that SLE is the initiating mechanism for excessive lateral vibrations on footbridges, including canopy walkways.

According to Dallard et al (2001), the dominant human induced force is the vertical force. Subsequently, it became the most discussed force in pedestrian bridge design guidelines (Setra, 2006; Heinemeyer et al., 2009). However, well-known examples such as the inauguration of the London Millennium Bridge in 2000 and the Solferino footbridge in Paris have shown that the lateral force is significant and may cause excessive lateral vibrations on a pedestrian bridge (Ingolfsson, et al., 2012). Surprisingly, even with the significant number of alarming reports concerning excessive lateral vibrations on footbridges, only a few have reported a structural collapse due to this force. Research and reports submitted after the vibration failure of the London Millennium Bridge indicated that this force was not only underestimated or not well understood, but it was also not unique to footbridges of a similar structural form to those mentioned above (Dallard, Fitzpatrick, et al., 2001b). Excessive lateral vibrations were also observed on other footbridges of different shapes, sizes and functions (Dallard, et al., 2001 & Ingolfsson, et al., 2012).

Chapter 1- Introduction

Itumeleng Ragoleka

In response to the lack of understanding of this previously underestimated phenomenon, an international conference focusing on the design and behaviour of footbridges was organised in Paris in 2002, known as "Footbridge 2002" (Caetano et al., 2009a). The conference attracted 69 academic paper contributions. This conference lead to the fib guide on design of footbridges in 2005 (Caetano et al., 2009a). Following this publication was the release of the widely-used guideline from the French road authority, Setra, in 2006 which mainly deals with the dynamic behaviour of footbridges, then the updating of the Eurocode and finally the publishing of the book entitled "Footbridge Vibration Design" in 2009 which integrates the contributions of a workshop arranged during the third international conference in 2008 (Caetano, et al., 2009 & Ingolfsson, et al., 2012). **Figure 1-1** affirms the increase in percentage of published papers pertaining to and addressing the problem of SLE since 1971 (Venuti & Bruno, 2009).

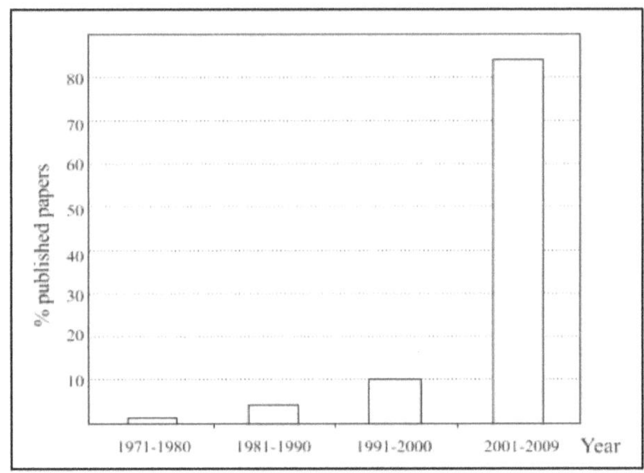

Figure 1-1: Percentage of published papers in the Civil Engineering field regarding SLE (Venuti & Bruno, 2009)

During a study of a crowd of pedestrians on a footbridge, video footage of pedestrians traversing on a laterally vibrating bridge deck revealed that pedestrians tend to synchronise their footsteps to the motion of the deck (Roberts, 2003 & Newland, 2004). Newland (2004) claims that the lateral sway of a pedestrian's centre of mass occurs at half the walking pace of the pedestrian. Considering that the average walking pace of a pedestrian is 2Hz, the centre of mass thus fluctuates at approximately 1Hz. Many footbridges have their first lateral natural frequency within the range of 0.8Hz to 1.2Hz and thus become prone to the effects of resonance when subjected to pedestrian lateral forces induced at these frequencies (Dallard, et al., 2001). The initial lateral vibrations of the bridge are caused by random lateral human induced forces. As soon as these vibrations increase and become perceptible to the pedestrian, the pedestrian finds comfort in synchronising his/her step with the motion of the bridge. Instead of counteracting the motion of the bridge, the synchronised motion of the pedestrian and bridge amplify the sway of the deck which leads to increased levels of discomfort or possible structural damage (Dallard, et al., 2001).

Studies on footbridge vibrations (Ingólfsson, Georgakis & Jönsson, 2012) have confirmed that the SLE phenomenon does not occur instantly, but gradually develops in proportion to the number of pedestrians on the footbridge. The London Millennium Bridge had approximately 200 pedestrians on it when lateral vibrations at 0.9Hz and an amplitude of 70mm was observed at the centre span (Newland, 2004). The steel arch bridge at Erlach in Germany was occupied by 300-400 pedestrians when excessive lateral vibrations at 1.1Hz occurred (Ingolfsson, et al., 2012). From such observations, researchers hypothesised that there probably exists a critical number of pedestrians who synchronise with the lateral motion of the bridge and trigger it to vibrate excessively beyond comfort and safety limits (Roberts, 2003; Newland, 2004 & Ingolfsson, et al., 2012).

1.2 Observed cases

Synchronous lateral excitation is not exclusive to any structural form or function and thus many footbridges subjected to this phenomenon have been reported globally (Ingólfsson, Georgakis & Jönsson, 2012). A few well known and distinct examples are discussed below.

The London Millennium Bridge is an example where SLE did not cause any structural damage but instead, compromised the safety and comfort of the users of the structure. The bridge is situated across the Thames River, connecting the City of London to St Pauls Cathedral. The official opening of the bridge was on the 10th of June 2000. On this day, around 80,000 to 100,000 pedestrians traversed the bridge with approximately 2,000 pedestrians on the bridge at any one time. This amounted to a maximum crowd density between 1.3 and 1.5 pedestrians per square meter. Synchronised lateral motion between the pedestrians and the bridge deck were observed at the south (108m) and centre (144m) spans of the bridge. Video footage showed amplitudes of 50mm at a frequency of 0.77Hz for the south span and 70mm at a frequency of 0.95Hz for the centre span. The latter mentioned span was observed to oscillate at this rate and magnitude when occupied by more than 200 people. Two days after this incident the London City Authority decided to close the bridge and investigate the cause of the vibrations to propose a suitable solution. (Dallard, et al., 2001)

Chapter 1- Introduction

Itumeleng Ragoleka

Another well-known example featured in similar discussions is the Toda Park Bridge in Japan. It is a cable-stayed bridge that connects a stadium and a bus terminal. It has an overall length of 179m divided into a main span of 134m and a side span of 45m. The frequency of the fundamental lateral vibration mode was between 0.9Hz and 1.0Hz. Shortly after the bridge was opened in 1989, a boat race was organised nearby resulting in over 20,000 pedestrians traversing the bridge. With approximately 2,000 pedestrians on the bridge at one moment and a crowd density of 2.1 ped/m^2, 20% of the pedestrians had their head motion synchronised to the lateral motion of the bridge. The bridge experienced a lateral vibration frequency of 0.9Hz, its natural frequency, and an amplitude of approximately 10mm. Although this might seem like an insignificant amplitude, reports confirm that many pedestrians felt uncomfortable and unsafe. (Ingolfsson, et al., 2012 & Hauksson, 2005)

Although there is a paucity of literature concerning the details of the Solferino Bridge in Paris, it is still regarded as a prominent example amongst those that exhibited unexpected lateral vibrations. It is a 140m long steel arch footbridge stretching over the Seine river in Paris. It was officially opened on the 15th of December 1999 and immediately experienced severe lateral oscillations. Following these observations, the bridge was closed for a comprehensive test program which continued for one year. During this period, many investigations were conducted, including vibration and crowd testing, which resulted in a solution to install 14TMDs and finally render the bridge open to the public in November 2000. (Ingolfsson, et al., 2012)

The Changi Mezzanine Bridge is a 140m long span steel footbridge located inside a tunnel at the Changi Airport, Singapore. The design of this bridge was completed prior to the vibration phenomenon observed on the London Millennium bridge and had fundamental frequencies in the lateral and vertical direction below 2Hz. Morever, the bridge was expected to operate with a normal maximum load of approximately 100-200 pedestrians. The consulting engineer on this project requested that an operational modal test be conducted on this bridge prior to its inauguration. The tests revealed a fundamental lateral vibration mode at approximately 0.9Hz and a fundamental torsional mode at approximately 1.64Hz. This suggested that the bridge could experience adverse synchronous lateral excitation due to a pedestrian walking pace of 1.8Hz. Similarly, the torsional mode could be triggered by the lower range of predominant footfall frequencies along with a relatively low modal mass. (Brownjohn, Fok, Roche & Moyo, 2004a)

Table 1-1 below, although not exhaustive, shows a list of footbridges which have shown evidence of excessive lateral vibration. This list also confirms the claim that the issue of excessive lateral excitation is not exclusive to a particular structural form but rather primarily dependent on the degree of human-structure interaction (HSI) (Newland, 2004a).

Table 1-1: Examples of reported cases of pedestrian-induced vibration (Fujino & Siringoringo, 2015)

Reported occurrences of pedestrian-induced vibration	Reported year	Bridge type	Observed vibration
Toda Park Bridge (Japan) (Fujino et al.1993)	1989	Cable stayed	Lateral
Pont de Solferino (France)(Dziuba et al.2001)	1999	Double-arch steel	Lateral
London Millennium Bridge (UK) (Dallard at al.2001)	2000	Suspension tension ribbon	Lateral
Lardal Bridge (Norway) (Ronnquist et al.2008)	2001	Arch timber	Lateral
Nasu Shiobara Bridge (Japan) (Nakamura 2003)	2002	Suspension	Lateral
Changi Mezzanine Bridge (Singapore) (Brownjohn et al.2004)	2002	Flat steel arch	Lateral & Torsion
Clifton Bridge (UK) (Macdonald 2008)	2003	Suspension	Lateral
New Coimbra (Portugal) (Butz et al.2008; Caetano et al.2010)	2006	Steel box	Lateral & Vertical
Well-am-Rhein (Germany) (Strobl et al.2007)	2007	Steel arch	Lateral

1.3 Objectives of this research

The key issue in this study leads to two core objectives:

(1)   The first core objective of this book is to determine the critical number of pedestrians required to trigger lateral synchronisation and lock-in on the Boomslang. The Arup model and the vibration comfort limit method will be used to determine this result.

(2)   The second core objective of this book is to attempt to elucidate the real dynamic mechanism(s), prior to synchronisation and lock-in, which initiate excessive lateral vibrations on a footbridge. The frequency component associated with the first lateral mode of vibration will be monitored in the time domain. The synchrosqueezed wavelet transform will be used to obtain this explanation.

Considering the unique design of canopy walkways, perceived vibrations can also be intentionally designed or built into the structural system in order to enhance the tourism experience. Currently, there is a lack of data on typical vibration levels characterizing these important structures. As a subobjective, this book seeks to collect data on the Boomslang Canopy walkway which can be used to inform the design of similar structures and for benchmarking the vibration performance of similar structures.

1.4 Scope and limitations of this research

This research will only review literature pertaining to pedestrian induced lateral vibrations on footbridges and canopy walkways. Primary attention will be devoted to discussing the walking gait and the lateral force induced by walking pedestrians on footbridges. The results and discussion thereof will be focussed on the lateral behaviour of the footbridge. The synchrosqueezed wavelet transform theory will also be reviewed in context of monitoring frequency data only. Only the Boomslang will be studied in this book. Related content of other footbridges will only be referenced as a matter of context and comparison.

1.5 Structure of this document

The remainder of this document will be structured in the following manner:

Chapter 2 will present the literature reviewed pertaining to pedestrian induced lateral forces on footbridges. This includes the fundamental concepts of pedestrian walking frequency, velocity and stride length. The idea of SLE, which lead to the development of the Arup stability criterion will be discussed. The final section of this chapter reviews the widely used design guidelines namely, Setra and HiVoSS. A comparison of these guidelines and the implications of the assumptions made in these guidelines will be discussed.

Chapter 3 is a continuation of the literature review which briefly reviews prominent signal processing techniques. A review on the theory of wavelet transforms is presented leading to the adopted synchrosqueezed wavelet transform.

Chapter 4 presents a detailed methodology followed in this study. The results obtained from the conducted investigations are presented and explained in Chapter 5. A coherent and contextualised discussion of the results is presented in Chapter 6. Thereafter, a wholistic conclusion with recommendations for further research are presented in Chapter 7.

Chapter 2

2 Literature review – Part 1

2.1 Introduction

Vibrations of footbridges are a prominent and pertinent issue in the current design of footbridges. Modern slender footbridge designs, including canopy walkways, aimed at improving aesthetics and capitalising on cost-effective methods of design have resulted in lightweight structures which have a high ratio of live to dead loading (Caetano, et al., 2009). The consequence of this trend has been the fact that many footbridges have become more susceptible to vibrations when subjected to dynamic loading events. Among many sources of dynamic loading, such as wind loading, the most common on footbridges are pedestrian induced footfall forces due to walking, jogging and jumping (Bachmann & Ammann, 1987).

Natural frequencies of modern slender footbridges tend to be low and this leads to excessive vibrations caused by pedestrians forces. These vibrations usually lead to serviceability problems which are validated by observing discomfort and emotional responses from pedestrians (Živanovi et al., 2005). Structural damage or collapse due to pedestrian induced excitation is rarely observed, however it cannot be completely eliminated as a possibility (Ingólfsson, Georgakis & Jönsson, 2012).

Research has shown that there is a relationship between the increase in pedestrian density on a bridge and the lateral amplification of the bridge deck, which leads to synchronisation and "lock-in" (Dallard, Fitzpatrick, et al., 2001b). The distinction between synchronisation and "lock-in" in this text is that synchronisation refers to the matching of one pedestrian's walking gait with another in the crowd whereas "lock-in" refers to the matching of the oscillating frequency of the bridge and the step frequency of the crowd (Brownjohn, Zivanovic & Pavic, 2008). The current contestation is around which event occurs first: an allusion to the chicken and egg scenario (Brownjohn, Zivanovic & Pavic, 2008). Despite this dispute, the common agreement among researchers in this field is that there is a critical number of pedestrians that trigger lateral synchronisation or lock-in and consequently propel the bridge beyond comfortable vibration limits. Research has also found that the critical number of pedestrians is unique to each structure and depends on the dimensions and dynamic properties of the structure (Dallard, Flint, et al., 2001).

2.2 Synchronous lateral excitation (SLE)

According to Fujino & Siringoringo (2015), the earliest reports of pedestrian-induced vibrations date back to the year 1850. Since then, pedestrian-induced vibrations, particularly lateral vibrations, have been observed on other footbridges such as the Toda Bridge in 1989 when a congested crowd traversed over the bridge. Minimal attention was given to researching this phenomenon until similar events were observed on the London Millennium bridge and the Solferino bridge.

As a result of the vibration serviceability failure of the London Millennium Bridge in the year 2000, extensive research effort was devoted to investigating the fundamental problem that triggered the excessive lateral vibrations witnessed on the bridge (Caetano, et al., 2009). Researchers, including Arup, concluded that the failure of the bridge was caused by the lateral synchronisation of the pedestrians who attended the inauguration event of the bridge and were excited to walk over the bridge for the first time (Dallard, Fitzpatrick, et al., 2001b).

Fujino et al. (1993) conducted full-scale tests and included video cameras to monitor the gait of pedestrians on the Toda-bridge in Japan (Živanović, Pavic & Reynolds, 2005; Nakamura & Kawasaki, 2006; Newland, 2004). From the video recording, Fujino et al. (1993) concluded that approximately 20% of the pedestrians were walking in phase with each other. Not only were the pedestrians walking in phase, but they were also walking in phase with the oscillations of the bridge deck. These were the first observed signs of lateral synchronisation and lock-in. Subsequently, it became important to investigate which factors instigated lateral synchronisation and lock-in.

An important factor found to initiate lateral synchronization is the high density of pedestrians on a bridge (Caetano et al., 2009a). Limited space causes pedestrians to synchronize their gait in an attempt to avoid "bumping" into one another. This was observed on the Toda-bridge crowded with approximately 2000 pedestrians which equated to 2.1p/m² (Ingólfsson, Georgakis & Jönsson, 2012; Nakamura & Kawasaki, 2006). Roberts (2003) claims that pedestrian densities of 1.3 and 1.5p/m² were observed on the London Millennium bridge on its inauguration day. However, Bachmann & Ammann (1987) reported that the maximum physically possible crowd density is between 1.6-

1.8p/m² and beyond this range it is inevitable that one pedestrian's movement will be influenced by that of the nearby pedestrian. The common observation is that as the density of the crowd increases, the walking pace decreases and the degree of synchronisation increases.

Synchronisation, inherently, is not the concern, but rather the frequency at which synchronisation occurs. Pedestrians normally walk at frequencies between 1.5 and 2Hz (Nakamura & Kawasaki, 2006; Caetano et al., 2009a). Literature (Bachmann & Ammann, 1987; Zivanovic, Pavic & Reynolds, 2005; Ingólfsson, Georgakis & Jönsson, 2012) has consistently stated that the fundamental component of the lateral force induced by pedestrians on the footbridge is due to their centre of gravity which oscillates at half their walking pace. This means that the oscillation occurs between 0.75 and 1Hz (Živanović, Pavic & Reynolds, 2005; Caetano et al., 2009). Modern light-weight footbridges (London Millennium Bridge, T-Bridge, Pont de Solferino) also have their lateral fundamental frequencies between the range of the pedestrians' lateral walking frequency. The correspondence of these frequencies causes resonance which results in the adverse dynamic response of the bridge (Dallard, Flint, et al., 2001).

Nakamura (2006) and Newland (2004) agree that random walking by pedestrians causes minimal vibrations on a footbridge with amplitudes up to 1mm. However, as more pedestrians crowd the bridge the acceleration of the deck gradually increases and consequently amplifies the oscillation range of the deck. Amplified accelerations of the bridge deck make pedestrians feel uncomfortable, cause loss of balance and eventually prevents them from continuing their journey. The desire for pedestrians to maintain their balance on a vibrating bridge deck causes them to spread their feet further apart and adjust their motion to that of the deck (Zivanovic, Pavic & Reynolds, 2005). This is known as the 'lock-in' effect. As more pedestrians synchronise their motion, the induced lateral force on the bridge amplifies and thus perpetuates the oscillating motion of the bridge. In such circumstances, reducing the number of people on the bridge, disrupting and/or stopping their movement is seemingly the only mitigation options for the problem (Zivanovic, Pavic & Reynolds, 2005).

Figure 2-1 and **Figure 2-2** show the results obtained from studies conducted to understand the probability of the lock-in effect occurring on a laterally vibrating bridge at a frequency of 0.75Hz or 0.95Hz (Newland, 2004b). The decision to tune the vibrating platform at these frequencies was strategic to obtain quasi-realistic results since the average lateral walking frequency of pedestrians is within this range. **Figure 2-1** demonstrates that there is a positive correlation between the amplitude of the platform and the probability of lock-in. It is shown that deck-amplitudes beyond 15mm increase the probability of 'lock-in'. Furthermore, **Figure 2-2** demonstrates a positive correlation between the amplitude of the platform and the dynamic force induced by the pedestrians on the bridge. The distinct conclusion to be made here is that it is vital to maintain low vibration levels and induced pedestrian forces to avoid lock-in.

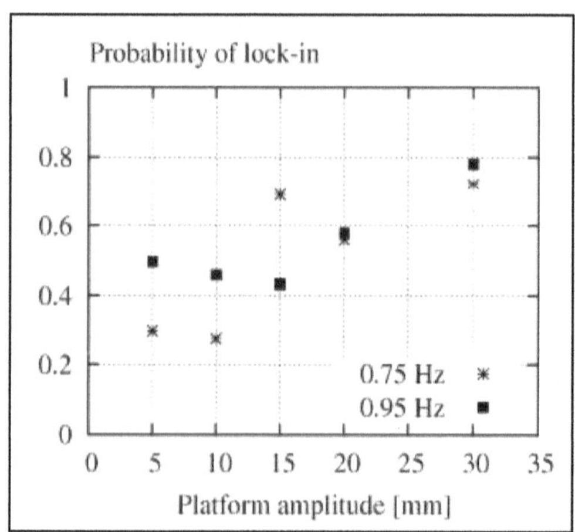

Figure 2-1: Relationship between the probability of lock-in and the platform amplitude (Dallard, Fitzpatrick, et al., 2001a)

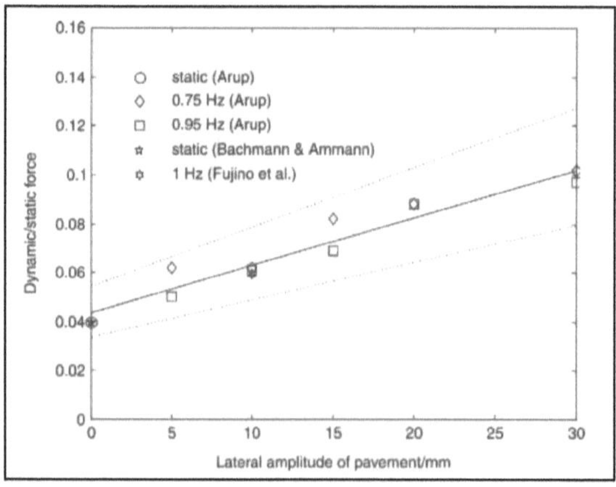

Figure 2-2: Measured values of pedestrian lateral dynamic force/static weight as functions of pavement amplitude (Newland, 2004.)

Having observed the detrimental lateral behaviour of the London Millennium Bridge, it became necessary for the scientific community to understand this phenomenon and eventually derive numerical formulae which would possibly assist with developing appropriate solutions for the issue. Some of the necessary numerical formulae developed covered the description of pedestrian forces and their interaction with the bridge structural system, also known as human-structure interaction (HSI) (Venuti & Bruno, 2009).

2.3 Pedestrian loading theories

2.3.1 Background

Single pedestrian motion on a rigid surface is the fundamental premise on which synchronised crowd dynamic load models are developed. The behaviour of a single pedestrian on a rigid surface is considered as unimpeded motion since it is not affected by neither the presence of other pedestrians nor by the vibration of the surface (Venuti & Bruno, 2009). Certain parameters are used to describe the free behaviour of the pedestrian and these are the walking velocity (v), the step frequency (f_v) and the step length (l_s). These parameters are related by the fundamental law: $v = f_v.l_s$. Venuti & Bruno (2009) state that the data concerning these parameters was sourced mainly from biomechanics and transportation research, but in recent years some results have also been published in the structural engineering field due to the interest in footbridge dynamic behaviour under pedestrian loading (Masani et al (2002); Venuti & Bruno (2007); Ingolfsson et al (2012)). Considering that much of the research involved in understanding these parameters encompasses other phenomena, such as running, jogging and jumping (Caetano, et al., 2009), this research will focus only on the walking activity, which is the predominant activity observed on the Boomslang Canopy walkway.

Researchers refer to the unimpeded gait of a single pedestrian as the *free speed* of the pedestrian (Venuti & Bruno, 2009). Extensive research effort has been directed to the statistical description of the pedestrian free speed as a random variable using a Probability Density Function (PDF). The general consensus is that the Gaussian PDF preferably agrees with the measured data. Venuti & Bruno (2009) published a table (**Table 2-1** below) reporting on selected PDF parameters obtained by different authors. The discrepancies noted in the mean values are due to the fact that they have been measured in different geographical areas and traffic situations. Pedersen & Frier (2010) state that the free speed of a pedestrian is influenced by physiological and psychological factors, such as biometric characteristics of the pedestrian (body weight, height, age, gender), travel purpose and the type of walking facility.

Table 2-1: Mean value and standard deviation of pedestrian free speed (Venuti & Bruno, 2009)

Author	v (m/s)	σ_v (m/s)	Geograhical area
Fruin (1971)	1.40	0.15	US
Hankin and Wright (1958)	1.60	n.a	US
Koushki (1988)	1.08	n.a	Saudi Arabia
Lam et al (1995)	1.19	0.26	Hong Kong
Older (1968)	1.30	0.3	UK
Pauls (1987)	1.25	n.a	US
Ricciardelli et al (2007)	1.41	0.224	Italy
Sanhaci and Kasperski (2005)	1.37	0.15	Germany
Sarkar and Janardhan (1997)	1.46	0.63	India
Tanariboon et al (1986)	1.23	n.a	Singapore
Virkler and Elayadath (1994)	1.22	n.a	US
Weidmann (1993)	1.34	n.a	n.a

Limited research has been conducted to obtain stride length results to compute pacing speed. Wheeler (1982) published results suggesting that the stride length is correlated with step frequency. The relationship observed in the results was not described using a mathematical formula, instead it was only presented graphically using a polynomial approximation as shown in **Figure 2-3**. Pedersen & Frier (2010) contributed by establishing a mathematical formula, **Equation 2-1**, which fits well with the suggestion made by Wheeler (1982).

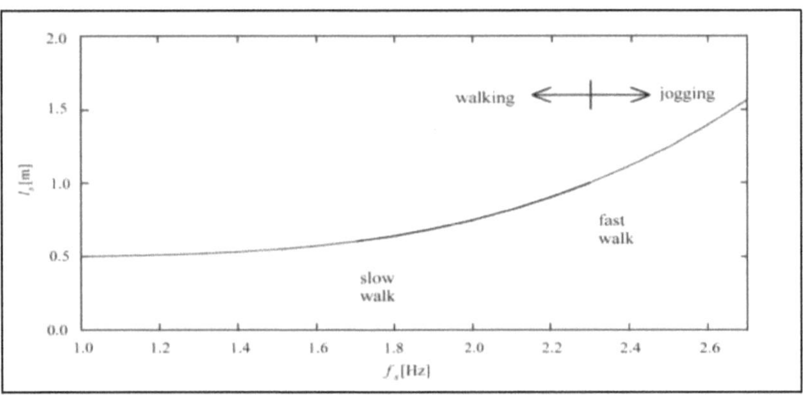

Figure 2-3: Relationship between step frequency and stride length in the 1 – 2.7 Hz range (Pedersen & Frier, 2010)

$l_s = 0.2011f^3 - 0.6021f^2 + 0.6462f + 0.2547$ (1<f<2.7Hz)	Equation 2-1

The final parameter is the step frequency which has been given much attention in studies relating to footbridge vibrations due to its influence on excitation and resonance. A distinction is made between the vertical step frequency and the lateral step frequency. The vertical step frequency is defined as the number of times a foot makes contact with the ground per unit of time (Venuti & Bruno, 2009). The lateral step frequency differs from this in that it is the number of times the same foot makes contact with the ground per unit of time, thus it is half of the vertical frequency (Venuti & Bruno, 2009). **Table 2-2** shows the results obtained by researchers regarding the vertical walking frequencies and the number of test volunteers involved in each research study. The variability in the results is due to physiological factors, psychological factors, geographical area and the type of surface used in the experiments. **Table 2-3** further distinguishes the type of walking by frequency mean. It is important to note that this book is particularly concerned with the "walking frequency" range as highlighted in **Table 2-3** and will thus focus on this range beyond this point.

Table 2-2: Mean value and standard deviation of the pedestrian walking frequency (Venuti & Bruno, 2009)

Author	f_s (Hz)	σ (Hz)	Sample (people)
Butz et al	1.84	0.126	n.a
Kerr and Bishop	1.9	n.a	40
Mastumoto et al	2	0.173	505
Pachi and Ji	2.0-1.83	0.135-0.11	800
Ricciardelli et al	1.835	0.172	116
Sanhaci and Kasperski	1.82	0.12	251
Zivanovic et al	1.87	0.186	1976

Table 2-3: Data on walking and running (Newland, 2004)

	Pacing frequency (Hz)	Forward speed (m/s)	Stride length (m)	Vertical fundamental frequency (Hz)	Horizontal fundamental frequency (Hz)
Slow walk	1.7	1.1	0.6	1.7	0.85
Normal walk	2	1.5	0.75	2	1
Fast walk	2.3	2.2	1	2.3	1.15
Slow running (jogging)	2.5	3.3	1.3	2.5	1.25
Fast running (sprinting)	3.2	5.5	1.75	3.2	1.6

Various equations describing the relationship between the walking velocity and the step frequency have been proposed. Similarly, various assumptions have been proposed as fitting with experimental measurements. Butz et al (2008) proposed $f = 0.7886 + 0.7868v$ as a suitable relationship, while Ricciardelli et al suggested $f = 0.024 + 0.754v$ and Bertram and Ruina proposed $f = 2.93v - 1.59v^2 + 0.35v^3$ (Venuti & Bruno, 2009). The latter formula, as seen in **Figure 2-4**, was forced to pass through the origin since as a valid hypothesis, the walking frequency is zero when the pedestrian is stationary.

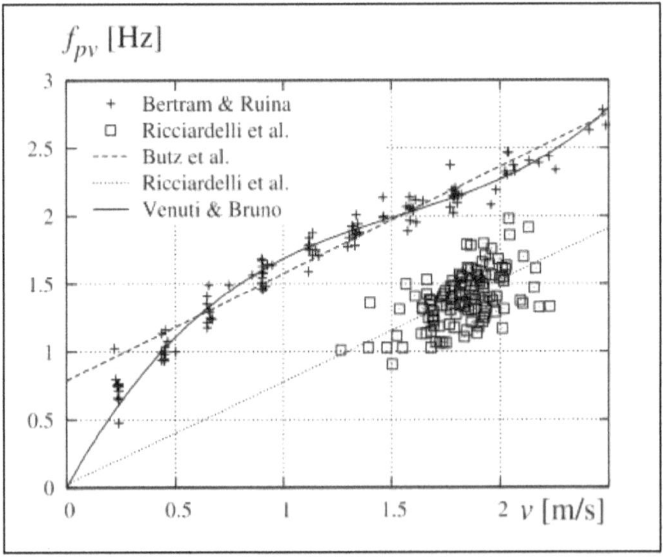

Figure 2-4: Examples of velocity and frequency relations (Venuti & Bruno, 2009)

The content in this background section seeked to recall the principal theories employed in the development of pedestrian force models. Moreover, it was important to review literature on the walking parameters since they are considered to be the primary contributors to the force exerted by pedestrians on a footbridge (Venuti & Bruno, 2009). The data collected has been of vital use for the development of pedestrian load models which eventually provide insight to the problem of synchronisation and lock-in.

2.3.2 Single pedestrian loading

Human-induced loading on a footbridge is dynamic and periodic by nature (Bachmann & Ammann, 1987). The loading is primarily known to be induced in the vertical direction at an average frequency of 2Hz. This component of the total force was measured to be 37% of the static weight of the pedestrian which is significantly larger than the lateral component of the same force (Newland, 2004c). However, since lateral vibrations of a bridge are the most intolerable to pedestrians, researchers have found it increasingly necessary to collect data and establish force models for the lateral component.

Newland (2004c) states that the lateral force component is approximately 4% of the static weight of the pedestrian and is applied at a mean of 1Hz. Furthermore, the shape (**Figure 2-5**) of the induced-lateral force depends on several parameters which are governed by intra and inter-subject variability. Intra-subjective variability is related to changes in the force from the same pedestrian, measured at two different time instances, whereas inter-subjective variability refers to the variability between people. Variations in the gait parameters during continuous walking is a form of intra-subject variability which causes random fluctuations in the shape of the force from each footstep (Ingólfsson, Georgakis & Jönsson, 2012). However, perfect periodicity in walking was assumed in order to allow the force time history of a series of consecutive footsteps to be modelled as a fourier series as shown in **Equation 2-2** (Brownjohn, Zivanovic & Pavic, 2008).

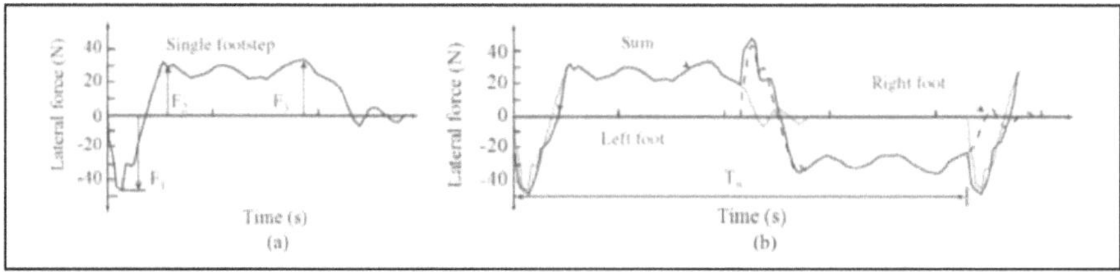

Figure 2-5: Typical shape of the lateral force induced by walking (a) single step and (b) a series of footsteps
(Ingólfsson, Georgakis & Jönsson, 2012)

$$F_l(t) = \sum_{i=1}^{\infty} G \propto_{i,l} \sin(2\pi i \frac{f_v}{2} t - \varphi_{i,l})$$	**Equation 2-2**

(where G is the pedestrian's weight (usually denoted as 700N), $\propto_i$ the Dynamic Load Factor (DLF) – i.e. the ratio of the peak dynamic force to the static weight of a pedestrian of the ith harmonic, φ_i - the phase shift of the ith harmonic and i is the order of the harmonic)

Bachmann & Ammann (1987) measured the first five lateral DLFs to be $\propto_i$ (i=1...5) = {0.039, 0.01, 0.043, 0.012, 0.015} (Zivanovic, Pavic & Reynolds, 2005). In later publications, Bachmann & Ammann (1987) suggested $DFL_1 = DFL_3 = 0.1$ for design purposes (Ingólfsson, Georgakis & Jönsson, 2012).

Many investigations conducted to quantify the forces induced by a single pedestrian on a footbridge were conducted on stationary platforms (Ingólfsson, Georgakis & Jönsson, 2012). Having hypothesized that the initial accelerations of the footbridge may have an effect on the induced lateral force of the pedestrian, novel research was conducted to understand the interaction between a pedestrian and a laterally moving platform with a particular interest in the effect of the moving platform on the lateral induced force (Ricciardelli & Pizzimenti, 2007). Since then, dynamic surface instruments have been introduced in investigations and since crowds are the real concern, streams of pedestrians have also been introduced to observe and understand the pedestrian-to-pedestrian and pedestrian-to-structure interaction.

2.3.3 Crowd loading

Crowds are considered to be extremely complex systems in their behaviour and interaction with a footbridge (Brownjohn, Zivanovic & Pavic, 2008; Venuti & Bruno, 2009). They represent collective behaviour which describes an individual's actions as being dominated by the influence of its neighbours, i.e. the individual behaves differently from how they would normally behave on their own (Venuti & Bruno, 2009).

Developing crowd loading models has been a substantial and vital task for the engineering community. The progression from single pedestrian loading to pedestrian stream loading and eventually interactive crowd loading has lead to improved and more realistic approximations of the observed dynamic behaviour of modern footbridges (Zivanovi, Pavic & Ingolfsson, 2010).

According to Fujino & Siringoringo (2015), Mastumoto et al. (1978) was among the first to propose a model that quantifies the force exerted by pedestrians on a bridge deck. In the model, the force per unit length [$F_p(t)$] exerted by pedestrians on the bridge [N_p] moving freely with their walking phases randomly distributed is expressed as:

$$F_p(t) = \frac{\sqrt{N_p} \propto g M_p}{L} \cos(f_p t)$$

Equation 2-3

(where M_p - mass of a single pedestrian; $\propto$ - DLF which depends on load direction and harmonic motion; L - footbridge span length; g - gravitational acceleration; and f_p – the dominant walking frequency)

This equation assumes that pedestrians move freely and unconstrained, which is more relevant in the case of a light stream of pedestrians. When a footbridge is crowded with a density of 0.6p/m^2 – 1p/m^2, pedestrians cease to move freely and eventually adjust their footsteps to match the motion of others. Hence, the formulation of a crowd-induced force(s) would be different from that describing a light stream of pedestrians (Fujino & Siringoringo, 2015).

Various investigations have proposed models for cases related to sparsely and very dense crowds. These models were developed on the basis of analytical and experimental studies encompassing models of bridges as stationary platforms, laterally moving platforms, or real pedestrian bridges (Fujino & Siringoringo, 2015). Therefore, assuming that pedestrians are distributed uniformly along the bridge length [L] and the phase angles of individual pedestrians are distributed uniformly over the period of walking, Roberts (2005) defined the lateral pedestrian load per unit length as:

$$F_p(t) = \frac{\sqrt{2N_p} M_p v_{ps} f_p^{\,2}}{L} \sin(f_p t)$$

Equation 2-4

(where v_{ps} – maximum amplitude of lateral motion; and M_p – mass of a single pedestrian)

Instrumented treadmills, either mounted on a stationary platform or one that vibrates laterally, have been introduced in many recent studies to verify the pedestrian-induced force. Experiments that use vibrating platforms are considered to be advantageous because moving platforms enable continuous vibration measurements and allow for more authentic walking of the test volunteers. Many of these experiments were conducted by researchers such as Ricciardelli & Pizzimenti (2007) and Ingólfsson et al. (2011).

Full-scale testing on real bridge structures to formulate pedestrian force models have also been explored. Following the London Millennium Bridge incident, Dallard et al. (2001) performed tests on this bridge which revealed that the dynamic force induced by pedestrians was approximately proportional to the bridge response. Furthermore, the force [F_p] per person was estimated to be proportional to the local velocity ($\dot{x}_{local}$) of the bridge deck by the lateral walking force coefficient k = 300Ns/m as $F_p(t) = k(\dot{x}_{local})(t)$ (Brownjohn, Zivanovic & Pavic, 2008; Fujino & Siringoringo, 2015). The local velocity is associated with the modal velocity ($\dot{x}$)(t) and bridge modal shape as ($\dot{x}_{local}$)(t) = $\emptyset$(x) ($\dot{x}$)(t). Therefore, the total modal lateral force exerted by [N_p] number of people distributed over the bridge can be calculated as (Fujino & Siringoringo, 2015):

$$F_p(t) = \sum_{i=1}^{N_p} \emptyset_i(x)^2 k\dot{x}(t)$$	**Equation 2-5**

Similarly, full-scale tests were conducted on the Solferino footbridge and it was concluded that the 'lock-in' phenomenon effectively occurred in the first lateral mode rather than in the torsional mode. Based on these results, the following pedestrian load intensity model was proposed:

$$F_p(t) = \frac{\gamma F_s \psi}{A} \sqrt{N_p} \cos(f_p t)$$	**Equation 2-6**

(where F_s – dynamic load amplitude of a single pedestrian; A – surface area of the bridge deck; and γ and ψ are functions of walking frequency and type of crowd. The load is represented in terms of an equivalent sinusoidal area load acting at the bridge's natural frequency [f_p])

An alternative approach observed in crowd loading formulations is the application of multiplication factors. Multiplication factors serve the purpose of amplifying the force of a single pedestrian and the result thereof is an equivalent load on a footbridge (Brownjohn, Zivanovic & Pavic, 2008).

Initially, research on crowd loading conducted by Matsumoto et al (1978) suggested estimating the vibration response of normal pedestrian traffic by multiplying the response to a single person exciting a footbridge at resonance by a factor $\sqrt{N} = \sqrt{\lambda T_0}$, where N is the number of people on the bridge at any time instant, while λ and T_0 represent the mean arrival rate (number of pedestrians per second) and the time required to cross the footbridge (expressed in seconds) (Brownjohn, Zivanovic & Pavic, 2008). This factor is derived by summing the responses of individual pedestrians arriving on the bridge according to the Poisson distribution and inducing the walking forces with equal frequencies and random phases. Dividing the factor by N can be interpretted as a synchronisation factor representing a portion of people in the crowd who, by chance, are walking in step with each other and render the effect of the rest of the crowd negligible (Brownjohn, Zivanovic & Pavic, 2008).

Brownjohn, Zivanovic & Pavic (2008) argue that the Matsumoto formula was relevent for footbridges vibrating predominantly in the vertical direction. An attempt to apply the formula on laterally vibrating footbridges, such as the T-bridge and the London Millennium bridge, resulted in significantly underestimated results for the T-bridge (a footbridge accomodating 2000 pedestrians, i.e. $\sqrt{2000}/2000 = 0.022$ which is ten times less than the appropriate factor of 0.2) (Fujino et al., 1993b). Moreover, the formula could not predict the vibration response of the London Millennium bridge (Dallard, Fitzpatrick, et al., 2001b), suggesting that the formula is not appropriate for footbridges prone to excessive lateral vibrations due to synchronized crowds (Brownjohn, Zivanovic & Pavic, 2008).

The possibility of synchronization among pedestrians in a crowd has posed a need to derive precise multiplication factors with which to amplify the single pedestrian load imposed on a laterally vibrating footbridge (Nakamura, 2004). Two expressions for the equivalent number of pedestrians (N_{eq}) have been established from emperical results using Monte Carlo simulations (Setra, 2006; Brownjohn, Zivanovic & Pavic, 2008). According to the HiVoSS and Setra guidelines, the N_{eq} for pedestrian densities less than $1p/m^2$ is $10.8\sqrt{N\xi}$ with an accompanying assumption stating that the pedestrians are free to move. For simulation purposes, this assumption leads to modeling random arrival times and normally distributed step frequencies centred around a natural frequency of the footbridge (Van Nimmen et al., 2014). The same guidelines express the N_{eq} for pedestrian densities beyond, and equal to, $1p/m^2$ as $1.85\sqrt{N}$ with an accompanying assumption that normal walking behaviour is obstructed, thus the simulation contains random arrival times but all pedestrians are now given the same step frequency (Van Nimmen et al., 2014).

Although there has been progress in developing mathematical descriptions of HSI pertaining to crowd loading, Shahabpoor, Pavic & Racic, (2017) acknowledge that these models are nowhere near being precise due to the complexities involved in crowd loading.

2.4 Stability criterion

Full-scale testing on real bridges lead engineers to the hypothesis that there is a threshold number of pedestrians required to trigger 'lock-in' and exacerbate the lateral vibrations of a pedestrian bridge (Dallard, Flint, et al., 2001). The concept of a stability criterion was soon associated with either a critical number of pedestrians or a critical acceleration response of the bridge (Fujino & Siringoringo, 2015). Both approaches (Setra, 2006; Dallard, Flint, et al., 2001) have reasonable and convincing arguments to support their usefulness in the research community.

Various stability criterions were proposed on the basis of different pedestrian excitation mechanisms (Fujino & Siringoringo, 2015). In the case of the direct resonance mechanism, which is the simplest mechanism for pedestrian-induced lateral vibrations, the assumption is made that vibrations are caused by pedestrians walking at the same frequency as the structure's lateral natural frequency. The stability criterion resulting from this condition is referred to as the "critical stability criteria" and describes an equivalent number of pedestrians whose walking frequencies are perfectly tuned to the structure's lateral frequency, thus causing 'lock-in'. Fujino et al., (1993) showed that approximately 20% of the pedestrians on the main span of the Toda bridge were synchronised to the lateral vibrations of the bridge, thus inducing a resonance force on the bridge. Additionally, Yoshida, Fujino & Sugiyama (2007) conducted similar investigations using advanced image processing techniques and detected an approximate head synchronisation of 50%.

Other stability criterion have been formulated on the basis of pedestrian-bridge dynamic interaction mechanisms. The pioneer of these is the Dallard, Flint, et al. (2001) model which models the lateral force exerted by the pedestrians on the bridge as a source of negative damping to the bridge's lateral motion. Therefore, the critical condition here is defined as the number of pedestrians beyond which the cumulative negative damping force becomes higher than the inherent damping force of the bridge. Based on this assumption, the Arup model was defined as:

$$N_{crit} = \dfrac{4\pi\xi M f_p}{\dfrac{k}{L}\int_0^L (\emptyset(x))^2\, dx}$$	**Equation 2-7**

(where $\emptyset(x)$ – mode shape; ξ – damping ratio; and f_p – frequency of the lateral mode of interest. M and L are the overall mass and length of the bridge)

Assuming an uneven distribution of pedestrians and a case where the maximum mode amplitude was normalised to unity, the generalised stability criterion was written as:

$$N_{crit} = \frac{8\pi\xi m_i f_i}{k}$$	**Equation 2-8**

(where m_i – generalized modal mass for fundamental lateral mode; and f_i – fundamental lateral frequency of the bridge)

Again, k is the lateral walking force proportionality factor with a value of 300Ns/m exclusively for the London Millennium Bridge. However, further investigations on other footbridges have been conducted and results have shown that the same "k" value produces satisfactory estimations of the "N" value (Caetano, Cunha, Moutinho & Magalhás, 2010) when the fundamental lateral frequency of the bridge is between 0.5Hz and 1Hz (Zivanovic, Pavic & Reynolds, 2005). Ultimately, the above critical criterion shows that low damping, low mass, or low frequency translates into a low critical value and therefore a higher risk for 'lock-in'.

Newland (2004) proposed a damping ratio criterion of the form (**Equation 2-9**) and a scruton number criterion of the form (**Equation 2-10**). Both of these were formed on relevant assumptions but have minute data to support their application.

$$\xi = \frac{\alpha\beta m}{2M}$$	**Equation 2-9**

(where ξ – modal damping ratio; α – ratio of movement of a person's centre of mass to movement of the pavement; β – correlation factor for individual people to synchronize with pavement movement; M – modal mass or, for a uniform deck, bridge mass per unit length; and m – modal mass of pedestrians or, for a uniform bridge deck with evenly spaced pedestrians, pedestrians mass per unit length)

$$S_{cp} = \frac{2\xi M}{m}$$	**Equation 2-10**

(where S_{cp} – pedestrian scruton number; ξ – modal damping ratio; M – modal mass or, for a uniform deck, bridge mass per unit length; and m – modal mass of pedestrians or, for a uniform bridge deck with evenly spaced pedestrians, pedestrians mass per unit length)

Setra (2006) proposed a strong argument suggesting that a critical acceleration is more intuitive as a stability criterion than the critical number of pedestrians since it corresponds to exactly what the pedestrians feel. The suggested threshold for lateral acceleration is 0.1-0.15m/s^2, and beyond this, the lock-in phenomenon is anticipated.

2.5 Comfort levels and accepted criteria

The functional use of footbridges is to assist pedestrians in their travel from one location to another in an area where obstacles are inevitable (Archbold et al., 2011). Since pedestrians are the primary users of footbridges, their needs are a priority to the designer. A critical need for a pedestrian is to feel safe and comfortable when walking over a footbridge.

The factors which contribute to the comfort of a pedestrian on a bridge can be divided into two categories: "soft" and "hardcore" factors. According to Butz et al., (2009), "soft" factors include;

- Number of people on the footbridge;
- Height above ground level;
- Orientation of the body;
- Transparency of the bridge deck;
- Expectancy of vibrations due to the appearance of the bridge;

while the "hardcore" factors include;

- Harmonic or transient excitation characteristics (frequency);
- Exposure time to vibrations or time spent on the bridge; and
- Severity of the bridge oscillations (amplitudes and accelerations)

This book focusses on one of the soft factors and all the hardcore factors mentioned above. The important soft factor in this book is the "number of people on the bridge". Venuti & Bruno (2009) mentions that the walking velocity is affected by the crowd density such that a higher crowd density results in a lower walking velocity. Consequently, the walking velocity affects the walking frequency which may eventually coincide with the natural frequency of the structure and cause resonance.

Large lateral amplitudes and peak accelerations reached during a vibration phenomenon are consequential to pedestrian comfort, synchronization and lock-in. Dallard, Fitzpatrick, et al (2001b) mention that the center span of the London Millennium bridge reached an amplitude of 70mm at 0.95Hz. According to Newland, (2004a), not only are such amplitudes detrimental to the structure due to enhanced lateral loading (as seen in **Figure 2-2**), but also increase the probability of synchronization and lock-in (as seen in **Figure 2-1**).

On the other hand, vibration comfort is a subjective matter because it strongly depends on the vibration direction, duration of exposure and the pedestrians' posture and activities (Matsumoto et al., 2010; Van Nimmen et al., 2014). Consequently, it is difficult to determine clear thresholds in relation to the comfort perceived by the pedestrians (Van Nimmen et al., 2014). The guidelines attend to this issue by presenting four intervals of acceleration levels with corresponding comfort levels, ranging from unacceptable vibration levels to maximum comfort which are presented in **Table 2-8** (Hivoss, 2007) and **Table 2-13** (Setra, 2006). Furthermore, as a matter of caution, the guidelines warn against the lock-in phenomenon which can be triggered by lateral accelerations beyond 0.1-0.15m/s^2 (Setra, 2006; Hivoss, 2007).

2.6 Structural damping of footbridges

The "lock-in" event as a phenomenon that perpetuates vibration serviceability issues has led to a growing demand for damping systems (Caetano, Cunha, Moutinho & Magalhães, 2010). Since synchronization and "lock-in" are phenomena that induce energy in the bridge system over time, it is imperative that the development of this energy is dissipated prior to the intiation of excessive lateral vibrations. It is for this reason that damping systems are used to control the vibration response of footbridges (Caetano et al., 2010).

Caetano et al., (2009) states that dynamic excitations on footbridges can be controlled by increasing the stiffness, damping and mass of the bridge. Other temporary measures involve limiting the number of pedestrians on the bridge to avoid large vibrations (Fujino & Siringoringo, 2015). It is important to note that the mere excitation of a footbridge is not a significant issue since it is rare that a structure will collapse due to mere dynamic excitation. However, the major issue is when the excitation of the bridge occurs at resonance frequency. In this situation, the amplitude of oscillation increases and renders the structure uncomfortable to use. **Figure 2-6** below demonstrates this idea by showing that the amplitude of the oscillations increases tremendously at resonance frequency (represented by β=1), notably when damping is non-existent (i.e. ξ=0) (Caprani, 2009). Therefore, it is vital to ensure that the approach used to dissipate the energy absorbed by the structure is effective and efficient (Caetano, Cunha, Moutinho & Magalhães, 2010).

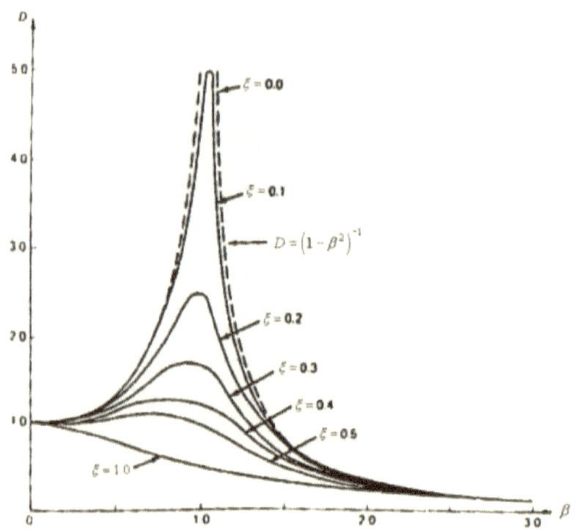

Figure 2-6: Resonance/damping curves (Caprani, 2009)

Modal damping ratios are a difficult parameter to obtain without conducting a full-scale dynamic test on a footbridge (Magalhães et al., 2010). At the design stage, the HiVoSS and Setra guidelines propose damping estimates according to the construction type as seen in **Table 2-6** and **Table 2-11**. These estimates were established from results obtained from investigations on many footbridges of various types and materials as part of the SYNPEX Project (Heinemeyer et al., 2009).

Alternatively, more accurate modal damping estimates can be obtained after the construction of the footbridge through experimental identification methods based on forced vibration, ambient vibration and free vibration tests (Magalhães et al., 2010). Forced vibration tests are considered to provide accurate results, however, they require very heavy and expensive equipment which has rendered them the unfavourable option of the three tests. Ambient vibration tests are considered to be very practical and economical. Natural frequencies and mode shapes obtained from them are accurate but the corresponding damping estimates present a significant scatter. Free vibration tests, based on the sudden release of a suspended mass or the sudden cut of a tensioned cable, are less practical than ambient vibration tests, but provide more accurate results (Magalhães et al.,

2010). Therefore, it is important to first consider the level of accuracy required for the damping estimates before choosing a test method.

Modal damping information informs vibration control solutions for foobridges (Heinemeyer et al., 2009). Instead of increasing the mass and stiffness of a footbridge, which may be considerably challenging, engineers opt for the employment of various types of damping systems such as tuned-mass-dampers (TMD's), visco-elastic-dampers and active control dampers. **Table 2-4** shows a list of footbridges which have experienced excessive vibrations due to pedestrian loading and the various damping systems installed to mitigate the prevailing issue (Fujino & Siringoringo, 2015). **Table 2-4** also shows that even though the bridges are of different types, the lateral vibration issues all occurred within 0.5-1Hz, which corresponds to the common lateral walking frequency range of pedestrians (Setra, 2006).

Table 2-4: **Examples of countermeasures for pedestrian-induced vibration using external damping devices** (Fujino & Siringoringo, 2015)

Bridge name (country)	Countermeasure	Controlled frequency (Hz)	Bridge type
Toda Park Bridge (Japan)	TLD (sloshing type)	L : 0.93	Cable stayed
Pont de Solferino (France)	1 TMD (L), 2 TMDs (V)	L: 0.81; V: 1.94, 2.22	Double-arch steel
London Millennium Bridge (U.K)	Viscous dampers and TMDs (L); vertical mass damper	L: 0.5, 0.8, 1.0; V: 1.2-2.0	Suspension tension ribbon
Stade de France (France)	TMDs (V)	V: 1.95	Girder
Forchjeim (germany)	1 semiactive TMD and MR damper	V: 1.0-3.0	Cable stayed
Bellagio to Bally (United States)	6 TMDs (V)	V: 1.7-2.2	Steel beam girder
New Coimbra (Portugal)	1 TMD (L), 6 TMD (V)	L: 0.85; V: 1.74-3.17	Shallow-arc girder

Note: V = vertical; L = lateral

2.7 Design guidelines for footbridges

The primary aim of footbridge design guidelines is to provide practical design procedures (Van Nimmen et al., 2014) and ensure secure and structurally sound footbridges which offer the appropriate levels of comfort to the pedestrian. To fulfil this mandate, these guidelines focus on establishing limits, informed by substantial research, to specific factors such as frequency, acceleration and pedestrian density (Caetano et al., 2009b).

The following section will review the methodology of the French Setra guideline and the European HiVoSS guideline. Both guidelines are widely used in engineering practice (Van Nimmen et al.,

2014). Furthermore, a comparison of the two guidelines will be provided with comments pertaining to the disparities observed in the guidelines.

2.7.1 European Design Guide (HiVoSS)

Footbridge classification:

The guideline classifies the footbridge according to the anticipated traffic which is a function of the location or purpose of the footbridge. A footbridge located in a rural environment is classified differently to a footbridge located in an urban environment and is thus analysed accordingly.

Table 2-5 shows five traffic classes varying from very weak traffic (TC 1) to exceptional dense traffic (TC 5). Each traffic class has an associated pedestrian density and the characteristics of each class are mentioned as a description of the level of freedom or restriction to the movement of the pedestrian.

Table 2-5: Traffic classes (Heinemeyer et al., 2009)

Traffic class	Density (P = Person)	Description	Characteristics
TC 1	15P/breadth*length	Very weak traffic	15 single persons
TC 2	$0.2\,P/m^2$	Weak traffic	Comfortable and free walking, Overtaking is possible, Single pedestrians can freely choose pace.
TC 3	$0.5\,P/m^2$	Dense traffic	Significantly dense traffic, Unrestricted walking, Overtaking can intermittently inhibit.
TC 4	$1.0\,P/m^2$	Very dense traffic	Freedom of movement is restricted, Uncomfortable situation, Obstructed walking, Overtaking is no longer possible.
TC 5	$1.5\,P/m^2$	Exceptional dense traffic	Very dense traffic and unpleasant walking, Crowding begins, One can no longer freely choose pace.

Equivalent load model:

Governing pedestrian load models are developed on the premise of a single pedestrian load model, however, footbridges are commonly subjected to forces due to groups of pedestrians or crowds. Therefore, a more realistic load model requires an amplification of the single pedestrian load and further account for the effects of intra-subject and inter-subject variability.

From the determined TC corresponding to a particular pedestrian density, the simplified load model consisting of an equivalent number (N_{eq}) of perfectly synchronized pedestrians is developed. The associated equation was derived from numerical simulations and was defined such that the same acceleration level was generated as the 95th percentile-value of the peak accelerations of 500 simulated streams of N random pedestrians.

For low pedestrian densities (d<1p/m²), free and unrestricted movement of the pedestrians is assumed. This results into random arrival times and normally distributed step frequencies centred around a natural frequency of the footbridge. The case for dense crowds (d> 1p/m²) assumes an obstruction to the normal forward movement of the crowd causing the stream of pedestrians to slow down. The simulation assumes random arrival times but all pedestrians are given the same step frequency. This results in a higher level of synchronization and a larger equivalent number of pedestrians. The guideline also states an upper limit value of 1.5p/m² beyond which the movement of pedestrians is considered to be impossible, thus significantly reducing the dynamic effects.

The equivalent number of pedestrians for low pedestrian densities (d<1p/m²) is defined as:

$$N_{eq} = 10.8\sqrt{N\xi}$$	**Equation 2-11**

while the equivalent number of pedestrians for dense crowds (d≥1p/m²) is defined as:

$$N_{eq} = 1.85\sqrt{N}$$	**Equation 2-12**

A uniformly distributed harmonic load that corresponds to the equivalent pedestrian stream is then defined as follows:

$$p(t) = G\cos(2\pi f_s t)\frac{N_{eq}}{S}\psi$$

Equation 2-13

(where G is the component of the force due to a single pedestrian and given as 280N for vertical vibrations and 35N for lateral vibrations, S is the area of the loaded surface (note that N=S×d), f_s is the walking frequency of the pedestrian or its harmonic multiple, and ψ is the reduction coefficient that accounts for the probability that the step frequency or its second harmonic approaches the considered natural frequency of the footbridge)

The reduction coefficient (ψ), as shown in **Figure 2-7**, has been determined on the basis of a statistical distribution of possible step frequencies, with a mean pacing rate typically around 2Hz, and its second harmonic. For lateral forces, the frequency range is divided by two owing to the particular nature of walking and the direction of the induced force.

The HiVoSS guideline distinguishes between two critical ranges of natural frequencies for which a calculation is required: Range 1 – possibility of resonance with the 1st harmonic and Range 2 – possibility of resonance with the 2nd harmonic (for vertical loading only).

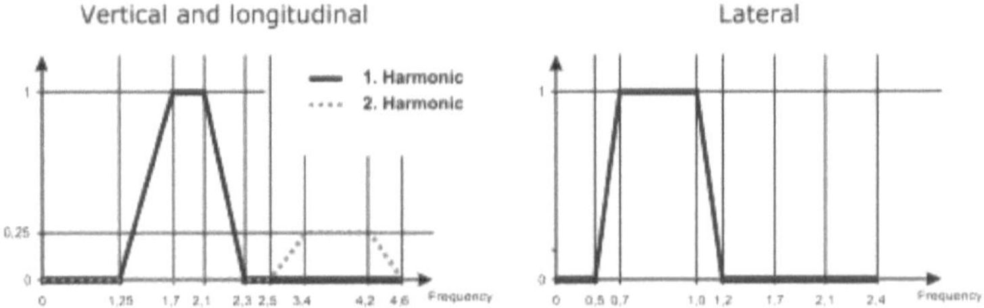

Figure 2-7: Reduction coefficient (ψ) (Heinemeyer et al., 2009)

Characterisation of the dynamic behaviour of the footbridge:

The guideline acknowledges that additional pedestrian mass may significantly alter the natural frequencies of the footbridge system. Single pedestrian and group loadings are usually negligible, however, streams of pedestrians or crowds may significantly decrease the natural frequency of a lightweight footbridge. The change in natural frequency may fall to a more or less critical frequency range for pedestrian induced dynamic excitation. Therefore, the guideline recommends that a check for the influence of additional modal mass be conducted if the additional modal mass exceeds 5% of the modal mass of the unoccupied bridge deck for the considered mode.

The damping ratio plays an important role in controlling the dynamic response of the structure at resonance conditions. This parameter can only be assumed at design stage and confirmed through full-scale dynamic testing after construction. The guideline suggests minimum and mean values of the damping ratio according to the considered construction type as shown in **Table 2-6**.

Table 2-6: Damping ratios according to construction type (Heinemeyer et al., 2009)

Construction type	Minimum ζ (%)	Average ζ (%)
Reinforcement concrete	0.8	1.3
Prestressed concrete	0.5	1.0
Composite steel-concrete	0.3	0.6
Steel	0.2	0.4
Timber	1.0	1.5
Stress-ribbon	0.7	1.0

Calculation of vibration levels:

The guideline outlines two methods of calculating the maximum acceleration of the footbridge due to various loading scenarios, namely: SDOF method and the Response Spectra Method.

The SDOF method operates on the premise that the dynamic behaviour of the structure can be described by a linear combination of serveral different harmonic oscillations exhibited by the structure. Therefore, the system can be decoupled into serveral different equivalent spring mass oscillators, each with a single degree of freedom. Each equivalent SDOF system (**Figure 2-8**) has one frequency and one mass equated to one natural frequency of the structure and the accompanying modal mass. The maximum acceleration a_{max} at resonance for the SDOF system is calculated as shown in **Equation 2-14**.

$$a_{max} = \frac{P^*}{m^*}\frac{1}{2\xi}$$	**Equation 2-14**

(where P* is the generalized load, m* is the modal mass and ξ is the structural damping ratio)

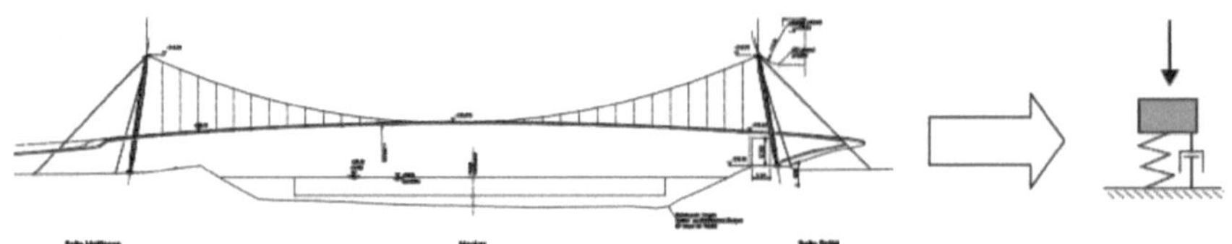

Figure 2-8: Equivalent SDOF oscillator for one natural frequency/ vibration mode of the structure (Heinemeyer et al., 2009)

The spectral design method devises a simple way to describe the stochastic loading and system response that provides design values with a specific confidence level. The assumptions of this evaluation are that the mean step frequency of the pedestrian stream coincides with the considered natural frequency of the bridge; the mass of the bridge is uniformly distributed; the mode shapes are sinusoidal; no modal coupling exists; and the structural behaviour is linear-elastic.

By analysis, the maximum acceleration is calculated for various pedestrian density situations as the 95th percentile of the maximum acceleration. **Equation 2-15**, along with **Table 2-7**, is used to calculate the maximum acceleration.

$$a_{max,d} = k_{a,d}\sigma_a$$	**Equation 2-15**

Table 2-7: Constants for lateral accelerations

d [P/m^2]	k_F	C	a_1	a_2	a_3	b_1	b_2	b_3	$k_{a,95\%}$
≤ 0,5		6,8	-0,08	0,50	0,085	0,005	-0,06	-1,005	3,77
1,0	$2,85\times10^{-4}$	7,9	-0,08	0,44	0,096	0,007	-0,071	-1,000	3,73
1,5		12,6	-0,07	0,31	0,120	0,009	-0,094	-1,020	3,63

Evaluation of vibration levels:

The comfort of the pedestrian is acknowledged as a subjective matter which strongly depends on the perceived vibrations. Furthermore, the perceived vibrations are dependent on the vibration direction, duration of exposure as well as the receiver's posture and activities. Consequently, it is difficult to determine clear thresholds in relation to comfort perceived by the pedestrian.

The guideline presents four intervals of acceleration levels with corresponding comfort levels, ranging from unacceptable vibration levels to maximum comfort. The ranges for lateral accelerations stipulated by the guideline are shown in **Table 2-8**.

Table 2-8: Define comfort classes with limit acceleration ranges (Heinemeyer & Feldmann, 2008)

Comfort level	Degree of comfort	Acceleration level horizontal/lateral (a_{limit})
CL 1	Maximum	$< 0.10 \, \text{m/s}^2$
CL 2	Medium	$0.1 - 0.30 \, \text{m/s}^2$
CL 3	Minimum	$0.30 - 0.80 \, \text{m/s}^2$
CL 4	Unacceptable discomfort	$> 0.80 \, \text{m/s}^2$

2.7.2 French Setra Code (2006)

Footbridge classification:

The Setra guideline classifies a footbridge according to the expected level of traffic on the footbridge. **Table 2-9** lists the possible traffic classes and the characteristics thereof which guide the designer and the client when classifying the footbridge. TC4 represents seldom usage of the bridge and the expectation of vibration problems is minimal. However, TC1 represents footbridges which are in constant utilization. High traffic volumes are expected on these footbridges and thus the probability of vibration problems is high.

Chapter 3- Literature review

Itumeleng Ragoleka

Table 2-9: Traffic class (Setra, 2006)

Traffic class	Characteristics
TC 4	Seldom used footbridge, built to link sparsely populated areas or to ensure continuity of the pedestrian footpath in motorway or express lane areas
TC 3	Footbridge for standard use, that may occasionally be crossed by large groups of people but that will never loaded throughout its bearing area
TC 2	Urban footbridge linking up populated areas, sujected to heavy traffic and that may occasionally be loaded throughout its bearing area
TC 1	Urban footbridge linking up high pedestrian density areas (for instance, nearby presence of a rail or underground station) or that is frequently used by dense crowds (demonstrations, tourists, etc), subjected to very heavy traffic

Equivalent load model:

The force of a single pedestrian walking at a constant speed v (m/s) along the centreline of the bridge deck can be represented as the product of the time component p(t) and a component describing its time-dependent position δ(x-vt):

$$P(x,t) = p(t)\delta(x - vt)$$	**Equation 2-16**

(where δ is the Dirac delta function and x the position of the pedestrian along the bridge centreline)

However, since footbridges are commonly subjected to multiple pedestrians, the guideline presents an equivalent pedestrian load adequate to produce accelerations in the 95th percentile. This equivalent load comprises an equivalent number of pedestrians (N_{eq}) who are assumed to be perfectly synchronized. Depending on the density of the pedestrians on the footbridge, the N_{eq} is determined accordingly.

The guideline makes a distinction between sparse and dense crowd conditions. Case 1 comprises TC3 and TC2 denoted by 0.5p/m² and 0.8 p/m² respectively. Subsequently, and similar to the HiVoSS, the N_{eq} is determined by **Equation 2-11** as these TC's are denoted by pedestrian densities below 1p/m².

Case 2 comprises a very dense crowd denoted by a pedestrian density of 1p/m². For this scenario, the N_{eq} is determined by **Equation 2-12**. Furthermore, the guideline makes provision for second harmonic effects stipulated in Case 3.

Similarly, the uniformly distributed harmonic load which corresponds to the equivalent pedestrian stream is then defined by **Equation 2-13**. However, the reduction factor in the Setra guideline differs from that presented by the HiVoSS as seen in **Figure 2-9**.

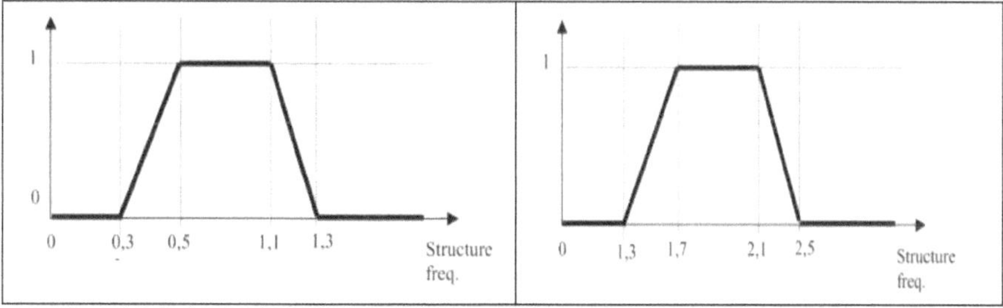

Figure 2-9: Reduction factor for lateral direction: 1st harmonic-left; 2nd harmonic-right. (Setra, 2006)

The guideline refers to 'ranges of risk' to describe the probability of resonance as seen in **Table 2-10**. Range 1 represents maximum risk of resonance and the degree of risk diminishes until Range 4 representing negligible risk of resonance. The colours are used to enhance intuition for the degree of risk, i.e. red is maximum risk and green is negligible risk.

Table 2-10: Frequency ranges for lateral vibrations (Setra, 2006)

Frequency (Hz)	0	0.3	0.5	1.1	1.3	2.5
Range 1						
Range 2						
Range 3						
Range 4						

Chapter 3- Literature review

Itumeleng Ragoleka

Characterisation of the dynamic behaviour of the footbridge:

The assessment of the influence of additional mass on the footbridge is conducted for scenarios pertaining to TC1-TC3. An examination of the natural frequencies of the structure is conducted for a scenario concerning an empty structure and one with a single pedestrian load of 700N/m². These results demarcate the upper limit and the lower limit of the frequency range.

The damping parameter also has a significant influence on the dynamic response of the footbridge. The Setra guideline suggests critical damping ratios for various construction types. **Table 2-11** shows the minimum and average critical damping ratios while **Table 2-12** shows the summarized critical damping ratios for various structural materials. It is observed that the average values in **Table 2-11** are the critical values in **Table 2-12** however, the timber value does not follow the same logic and the difference has not been explained by the authors.

Table 2-11: Minimum and average critical damping ratios for various structural materials (Setra, 2006)

Type of deck	Critical damping ratio	
	Minimum value (%)	Average value (%)
Reinforcement concrete	0.8	1.3
Prestressed concrete	0.5	1.0
Metal	0.2	0.4
Mixed	0.3	0.6
Timber	1.5	3.0

Table 2-12: Critical damping ratios for various structural materials (Setra, 2006)

Type	Critical damping ratio (%)
Reinforced concrete	1.3
Pre-stressed concrete	1.0
Mixed	0.6
Steel	0.4
Timber	1.0

Calculation of vibration levels:

In calculating the maximum vibration levels of the footbridge, the guideline uses the SDOF approach as shown in **Equation 2-17** (Gheitasi et al., 2016).

$$a_{max} = \frac{1}{2\xi}\frac{4|p(t)|w_d}{p_s\pi}$$	**Equation 2-17**

(where $|p(t)|$ is the magnitude of the applied load $p(t)$; w_d is the available width of the deck for pedestrian walking; and p_s is the total linear density, which is calculated as the sum of the linear density of deck and that of the pedestrians)

Evaluation of vibration levels:

The guideline deligates the responsibility of determining the comfort level of the footbridge to the owner. Four distinct comfort levels are established: maximum comfort, average comfort, minimum comfort and uncomfortable level. The guideline states that the concept of comfort due to the dynamic response of the structure is highly subjective making it difficult to establish vibration thresholds. Therefore, **Table 2-13** was adopted, using acceleration ranges, to distinguish the different comfort levels. The maximum comfort level is experienced between 0m/s^2 and 0.15m/s^2, however, caution is made for accelerations between 0.1m/s^2 and 0.15m/s^2 as this marks the trigger range for 'lock-in' (indicated by red line in **Table 2-13**). Again, the level of comfort diminishes with increasing acceleration as shown and emphasized by the colours in **Table 2-13**.

Table 2-13: Acceleration range for lateral vibrations. The acceleration is limited in any case to 0.1m/s^2 to avoid lock-in effect.(Setra, 2006)

Acceleration (m/s^2)	0	0.1	0.15	0.3	0.8
Range 1	Max				
Range 2			Mean		
Range 3				Min	
Range 4					

2.7.3 Comparison

The methodologies of the Setra and the European guideline are similar (Van Nimmen et al., 2014), however, minute differences were detected which may have an influence on the accuracy of the obtained results and the quality of decision making for a particular problem. Firstly, both guidelines agree that there are different crowd cases which cause distinct dynamic responses on a footbridge. The guidelines classify these in different traffic classes. In the HiVOSS guideline, the five traffic classes are defined by distinct pedestrian densities which correspond to the relative freedom experienced by pedestrians on the footbridge. On the contrary, the Setra guideline only provides a qualitative description, referring to the relative freedom experienced by the pedestrians, for each of its four traffic classes. Futhermore, it is mentioned in the Setra guideline that TC 1 to TC 3 should be evaluated with 2 cases: empty bridge and $700N/m^2$. Therefore, depending on the guideline used, the resulting equivalent loads will be different.

Both guidelines affirm the need to categorise different comfort levels. They both describe 4 comfort level classes: Maximum, medium, minimum and unacceptable level of comfort. The distinction between one class and another is established by an acceleration range. The HiVoSS guideline associates the maximum comfort level with lateral accelerations less than, but not equal to $0.1m/s^2$. However, the Setra guideline associates the same level of comfort with lateral accelerations less than $0.15m/s^2$. There is no academic evidence of whether there is a considerable change in experienced comfort within the range of 0.1-$0.15m/s^2$, however, both guidelines highlight the fact that lock-in is highly probable in this range. Furthermore, both guidelines agree on the acceleration ranges describing the remaining comfort level classes.

Lastly, a unique insight provided by the Setra guideline is the "risk of resonance" estimate. Four ranges of risk are highlighted (Range 1 = high risk to Range 4 = no risk at all) and they are each associated with a particular lateral frequency range. Although this information is beneficial for a designer or a researcher, the European guideline does not overtly state the risk of resonance in this manner.

Chapter 3

3   Literature review – Part 2

3.1 Signal processing techniques: A brief history

Modal parameter identification (MPI) of civil infrastructure through signal processing of vibration based data has been an academic pursuit in the fields of numerical model updating, vibration control, structural health monitoring and condition assessment (Perez-ramirez et al., 2016). The accurate identification of modal parameters of a civil structure largely depends on the excitation source for vibration behaviour of the structure. In cases were ambient vibration methods have been used, the obtained data usually exhibits non-stationary properties and contains considerable levels of noise. Therefore, a selected signal processing technique would need to overcome these challenges to provide informative results.

The Fast Fourier Transform (FFT) is the oldest and most utilized technique for signal processing in modal parameter identification tasks (Amezquita-Sanchez & Adeli, 2016). This technique transforms a time domain signal [x(t)] into the frequency domain and thus shows the frequency composition of the signal. The signal is estimated by a weighted sum of a series of sine and cosine functions (Gao & Yan, 2011). Its applications are seen in the use of the Peak Picking (PP) method which appears in papers such as Ren et al. (2004) and Ren & Peng (2005). The advantages of the FFT include it being relatively easy to implement and efficient for analyzing stationary and low noise level signals (Amezquita-Sanchez & Adeli, 2016). However, the FFT cannot depict spectral changes over time which is important when analyzing non-stationary signals (Amezquita-Sanchez & Adeli, 2016).

$$X(f) = \int_{-\infty}^{\infty} x(t) * e^{-2j\pi f t} dt$$	**Equation 3-1**

Acknowledging the shortcomings of the FFT, Gabor (1946) introduced the Short Time Fourier Transform (STFT) (Gao & Yan, 2011). The STFT divides the signal into small time windows which are analyzed using a FFT. This particular feature enables the STFT to efficiently analyze non-stationary signals and detect sudden changes in the frequency content of the signal (Amezquita-

Sanchez & Adeli, 2016). **Equation 3-2** shows the mathematical description of the STFT while results in Brownjohn et al., (2004) show a practical application of the STFT, also known as the spectrogram. The limitation of the STFT is the tradeoff between time and frequency resolution. Selecting a long analysis window results in good frequency resolution but poor time resolution. A short analysis window results in good time resolution and poor frequency resolution. This limitation is explained well by the Heisenberg uncertainty principle (Li & Liang, 2012). The concern about selecting an appropriately sized window to enhance the resolution of both time and frequency components is serious when analyzing a transient signal. Also, the STFT is unreliable when analyzing signals with closely spaced dominant frequencies (i.e closely spaced modes) (Amezquita-Sanchez & Adeli, 2016).

$$S(\tau,f) = \int x(t)w^*(t-\tau)e^{-2j\pi ft}dt$$	**Equation 3-2**

Considering the abundance of nonstationary signals in practice, the wavelet transform, shown in **Equation 3-3** was developed to mitigate the issues experienced by previously mentioned signal processing techniques. The wavelet transform provides a time-frequency representation of the signal using variable sized analysis windows. Various types of window functions, also known as mother functions, have been developed with unique properties which have unique effects in the signal analysis process. The different types of mother wavelet functions include the Haar, Daubechies, Mexican hat, Gauss, Morlet, Shannon, Synlets, Coiflets, Meyer, Spline and Gabor function which are illustrated in **Figure 3-1** below (Amezquita-Sanchez & Adeli, 2016). These functions are translated and dilated to obtain the approximations and detailed coefficients of the wavelet transform. The main advantages of the wavelet transform are its computational efficiency, data compression and noise elimination capabilities. For these reasons, the wavelet transform has gained much attraction in signal processing activities, however, it also has a few drawbacks which have lead to many iterations and modifications being pursued (Li & Liang, 2012).

$$WT(\tau,s) = \frac{1}{\sqrt{s}} \int x(t)\psi^*\left(\frac{t-\tau}{s}\right)dt$$	**Equation 3-3**

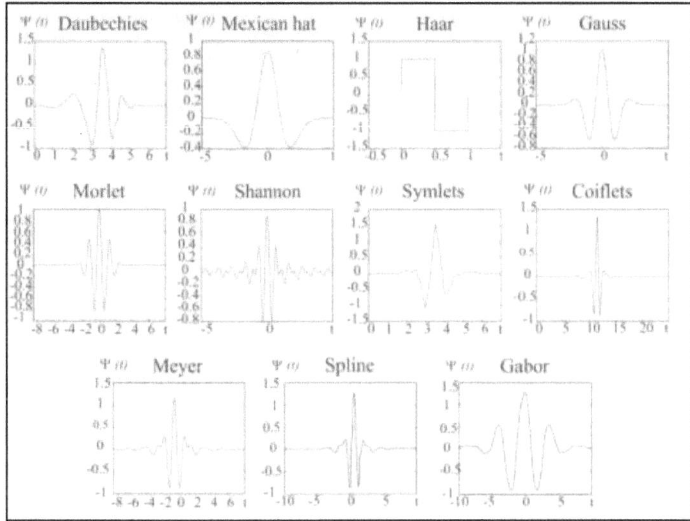

Figure 3-1: Mother wavelet functions (Amezquita-Sanchez & Adeli, 2016)

3.2 Synchrosqueezed wavelet transform

A well accepted and developed wavelet transform technique was proposed by Ingrid Daubechies (Daubechies, Lu & Wu, 2009) and was named the synchrosqueezed wavelet transform (SWT). The SWT provides a more accurate time-frequency representation compared to previously mentioned signal processing techniques. The SWT is also capable of providing superior results while analyzing noisy signals (Perez-ramirez et al., 2016). The SWT is also known as a reallocation technique due to its inherent ability of reallocating the continuous wavelet transform (CWT) coefficients based on the frequency information in order to obtain a clearer representation in both time and frequency domains (Amezquita-Sanchez & Adeli, 2016).

The SWT has been applied in fields of seismic data analysis (Wang, Gao & Wang, 2014), mechanical engineering (Feng, Chen & Liang, 2015) and modal parameter identification of civil engineering structures (Perez-ramirez et al., 2016). Regarding modal parameter identification of bridges, Perez-ramirez et al., (2016) shows the superior capability of the SWT in producing clear natural frequencies and damping ratio results with respect to time from ambient vibration data. However, minimal literature is available on the use of this technique for other bridge vibration data, particularly pedestrian bridges with low and closely spaced modes.

The analytical theory of the synchrosqueezed wavelet transform is documented in Daubechies, Lu & Wu, (2009). Herein, the methodology is described as follows: a purely harmonic signal of the form in **Equation 3-4** is assumed.

$$s(t) = A\cos(wt)$$	**Equation 3-4**

A mother wavelet (ψ) is selected and assumed to be concentrated on the positive frequency axis: $\hat{\psi}(\xi) = 0$ for $\xi < 0$. Therefore the continuous wavelet transform $W_s\,(a,b)$ of $s(t)$ will be

| $W_s\,(a,b)$ | $$\begin{aligned} &= \int s(t)\, a^{-\frac{1}{2}}\, \overline{\psi\left(\frac{t-b}{a}\right)}\, dt \\ &= \frac{1}{2\pi}\int s(\xi)\, a^{1/2}\, \overline{\hat{\psi}(a\xi)}\, e^{ib\xi}\, d\xi \\ &= \frac{A}{4\pi}\int [\delta(\xi - \omega) + \delta(\xi + \omega)]\, a^{1/2}\, \overline{\hat{\psi}(a\xi)}\, e^{ib\xi}\, d\xi \\ &= \frac{A}{4\pi}\, a^{1/2}\overline{\hat{\psi}(a\omega)}\, e^{ib\omega} \end{aligned}$$ | **Equation 3-5** |

Assuming that $\hat{\psi}(\xi)$ is concentrated around $\xi = \omega_0$, then $W_s\,(a,b)$ will be concentrated around $a = \omega_0/\omega$. However, the wavelet transform $W_s(a,b)$ will be spread out over a region around the horizontal line $a = \omega_0/\omega$ on the time-scale plane. Furthermore, an observation made by Daubechies is that although $W_s\,(a,b)$ is spread out in a, its oscillatory behaviour in b points to the original frequency ω, regardless of the value of a.

A suggestion was made to compute, for any (a,b) for which $W_s(a,b) \neq 0$, a candidate instantaneous frequency $\omega(a,b)$ using **Equation 3-6**.

$$\omega(a,b) = -i\big(W_s\,(a,b)\big)^{-1}\frac{\partial}{\partial b}W_s\,(a,b)$$	**Equation 3-6**

For the purely harmonic signal s(t), the result of $\omega(a,b)$ is ω as shown in **Figure 3-2**.

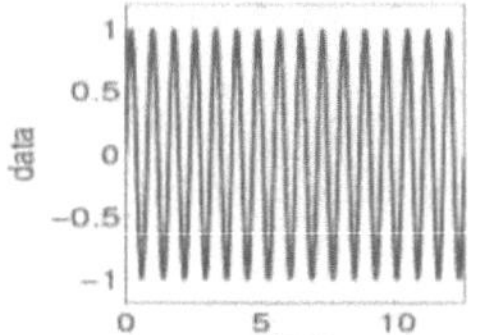 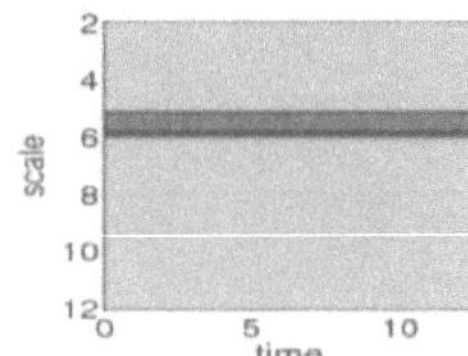 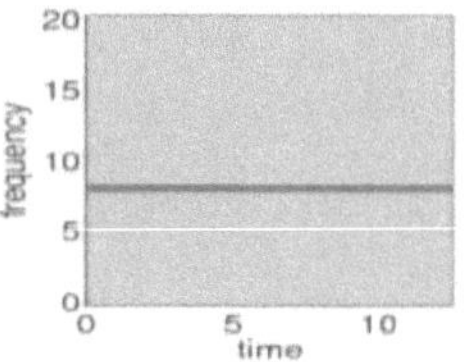

Figure 3-2: Left: The harmonic signal f(t) = sin(8t); Middle: The continuous wavelet transform of *f*; Right: Synchrosqueezed transform of *f*. (Daubechies, Lu & Wu, 2009)

The following step involves transferring the time-scale information to the time-frequency plane, according to the map $(b,a) \rightarrow (b, \omega(a,b))$. Another operation of synchrosqueezing involves 'binning' the frequency variable (ω) and the scale variable a. The result thereof was $W_s(a,b)$ computed only at discrete values a_k, with $a_k - a_{k-1} = (\Delta a)_k$ and its synchrosqueezed transform $T_s(\omega,b)$ was likewise determined only at the centers ω_l of the successive bins $[\omega_l - \frac{1}{2}\Delta\omega, \omega_l + \frac{1}{2}\Delta\omega]$, with $\omega_l - \omega_{l-1} = \Delta\omega$, by summing different contributions:

$$T_s(\omega_l, b) = (\Delta\omega)^{-1} \sum_{ak:|\omega(ak,b)-\omega l|\leq\Delta\omega/2} W_s(a_k, b)a_k^{-\frac{3}{2}}(\Delta a)_k$$	**Equation 3-7**

Another advantage of the synchrosqueezing method is its ability to reconstruct the original signal. The following argument shows how the signal is reconstructed:

$$W_s(a,b)a^{\frac{-3}{2}}da$$	$$\begin{aligned} &= \frac{1}{2\pi}\int_{-\infty}^{\infty}\int_0^{\infty} \hat{s}(\xi)\overline{\hat{\psi}(a\xi)}e^{ib\xi}a^{-1}dad\xi \\ &= \frac{1}{2\pi}\int_0^{\infty}\int_0^{\infty} \hat{s}(\xi)\overline{\hat{\psi}(a\xi)}e^{ib\xi}a^{-1}dad\xi \\ &= \int_0^{\infty}\overline{\hat{\psi}(\xi)}\frac{d\xi}{\xi} \cdot \frac{1}{2\pi}\int_0^{\infty}\hat{s}(\zeta)e^{ib\zeta}d\zeta \end{aligned}$$	

Setting $C_\psi = 2\int_0^{\infty}\overline{\hat{\psi}(\xi)}\frac{d\xi}{\xi}$, we then obtain (assuming that s is real, so that $\hat{s}(\xi) = \overline{\hat{s}(\xi)}$, hence $s(b) = (4\pi)^{-1}Re[\int_0^{\infty}\hat{s}(\xi)e^{ib\xi}d\xi]$)

$$s(b)$$	$$= Re[C_\psi^{-1}\int_0^{\infty} W_s(a,b)a^{\frac{-3}{2}}da]$$	

In the piecewise constant approximation corresponding to the binning in a, this becomes

$s(b)$	$\approx Re[C_\psi^{-1} \sum_k W_s(a_k, b)a_k^{\frac{-3}{2}} (\Delta a)_k]$ $= Re[C_\psi^{-1} \sum_l T_s(\omega_l, b) (\Delta\omega)]$	

Chapter 4

4 Methodology

The methodology presented below proposes a six (6) step process (as shown in **Figure 4-1**) designed to fulfill the above-mentioned objectives of this research. Considering the fact that the critical number of laterally synchronized pedestrians required to trigger lock-in on a footbridge is dependent on the modal parameters of the particular structure, firstly a finite element model of the Boomslang Canopy walkway will be developed to estimate the natural frequencies of the footbridge. Thereafter, ambient vibration testing will be conducted to collect vibration data to be used for modal parameter estimation. Modal analysis with ME Scope (approved vibration analysis software) will be performed. The Arup model will be used to evaluate the critical number of pedestrians required to trigger lock-in. Pedestrian interaction tests will be performed to assess the influence of pedestrian loading on the modal parameters of the bridge and the overall dynamic response of the bridge. Vibration levels of the bridge will be evaluated according to the established guidelines using Matlab 2016a. Furthermore, the critical number of pedestrians will be evaluated as per the vibration comfort limit method established by the design guidelines. Lastly, the synchrosqueezed wavelet transform will be used to either support or challenge the common notion that SLE is the initiating mechanism for excessive lateral vibrations on footbridges.

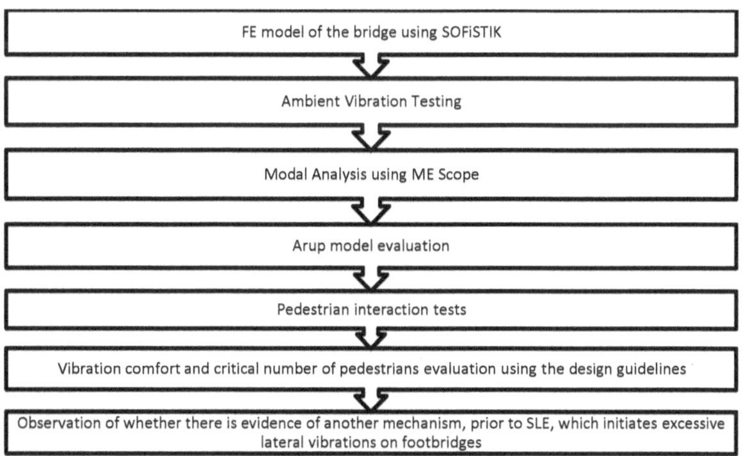

Figure 4-1: Methodology process

4.1 Description of Boomslang Canopy Walkway

The Boomslang Canopy walkway (**Figure 4-2**) is located in SANBI Kirstenbosch Gardens which is on the east side of Table Mountain in Cape Town. By client specifications, the structure needed to provide pedestrians (including wheelchair users) with a new experience of the garden and access to the canopy of the forest. The structure was also limited to causing negligible damage to the surrounding trees. For this reason, steel was the best material option to fulfil these requirements (SAISC Projects, 2015).

Figure 4-2: The Boomslang Canopy Walkway

Chapter 4 – Methodology

Itumeleng Ragoleka

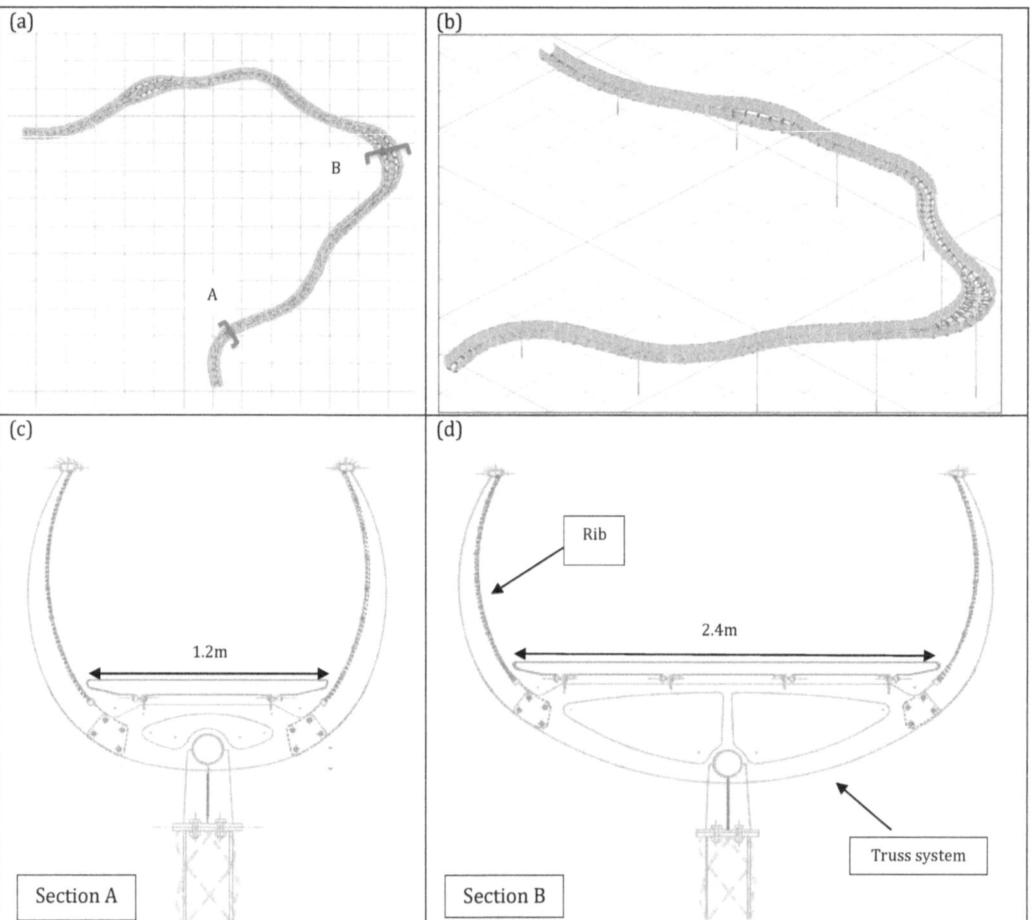

Figure 4-3: Boomslang Canopy walkway- (a) plan view; (b) Isometric view; (c) Section A; (d) Section B.

The footbridge is a multi-curve structure with a total deck length of 126m. The width of the bridge is mainly 1.2m (**Figure 4-3-C**) with two widened sections of 2.4m (**Figure 4-3-D**). The footbridge has 11 spans and a total dead weight of 2200kN.

The main spine of the walkway is tubular steel, with welded ribs and a light mesh providing the cross bracing. This allows the whole form to act as a bridge-spanning beam. The spine of the structure forms the bottom chord of a truss. The box section handrails complete the top chords of the truss system.

Ribs cut from a 8mm plate at 1m centres, serve both as stanchions and as the vertical elements of the truss system. These ribs have three segments bolted together below the deck. This ensured efficient use of material.

Two longitudinal angle rails, onto which the transverse walkway planking is fixed, also serve more than one purpose. During construction, when only the lower central portion of the walkway was erected, these angles served as top chord members of a triangular truss, with the circular hollow section being the bottom chord. At completion, they served both as chord members of the truss and load transfer elements from the walkway planking to the spine.

The 8mm rods forming the safety mesh contribute to the stiffness of the structure as truss diagonals. Their gradient varies with span, being steeper near the columns where shear forces are highest and shallower at mid-span. The curves soften the appearance of the walkway and give the structure an organic feel.

H-section steel columns are of various heights, gradually raising the walkway to a maximum of 12m above the ground at the approximate centre of the bridge. The columns are bolted on concrete foundations, providing translation stability, but not rotation stability. Lastly, the concrete abutments provide both translational and rotational stability in all directions.

4.2 Finite element model

According to Zivanovi, Pavic & Reynolds (2006), it is considered good practice to develop a reasonably detailed FE model of the as-built structure prior to conducting experimental tests. This assists with obtaining estimated natural frequencies and mode shapes. Furthermore, this will provide insight into the dynamic behaviour of the canopy walkway which assists with full-scale test planning and preparation.

The author developed a FE model of the Boomslang footbridge using the SOFISTIK SSD 2018 package. This was accomplished through the use of a 3D AutoCad drawing provided by the design engineer, Mr Henry Fagan of Henry Fagan & Partners in Cape Town, South Africa. Furthermore, to complete the model, good engineering judgement was employed to compensate for the limited information obtained from the technical drawings.

The model was developed by creating multiple elements as listed in **Table 4-1** below.

Table 4-1: Element modelling

Cross-section	Model Type	Dimensions
Deck	Glue laminated timber	76mm thickness
Main Spine	Tube	165mm Øin and 6mm thickness
Hand Rail	Rolled Steel > SH	40×80×4 (mm)
Longitudinal angle rail (left)	Rolled Steel > L	80×60×6 (mm)
Longitudinal angle rail (right)	Rolled Steel > L	60×80×6 (mm)
Column	Rolled Steel > HEA	100 (mm) flange
Stanchions	Rolled Steel > BAR EN10060	Ø30mm
Ribs	Rigid links	N/A

The dynamic behaviour of the bridge was evaluated using the Eigenvalues task in SOFISTIK (SSD) 2018. There are two computational methods prescribed by SOFISTIK to compute the dynamic behaviour of the developed model which are the ASE and DYNA modules. The DYNA module was chosen as the computational method in this project. As commonly advised for footbridge evaluations, eigenvalues up to the 4th harmonic (Reynolds, 2014) should be evaluated and thus sufficient eigenvalues were computed to capture lateral modes up to 5Hz. In the solver settings, the Lanczos method was employed to compute the dynamic characteristics of the bridge.

The results of this analysis, showing natural frequencies and mode shapes, are reported in section 5.1 of the results chapter.

4.3 Ambient vibration tests

The necessity of modal testing is due to the awareness of the shortcomings of FE models in predicting the real vibration behaviour and the operational modal parameters of a structure (Ren et al., 2004). Operational modal parameters of a footbridge can be obtained by conducting forced vibration tests or ambient vibration tests (Caetano et al., 2010). Forced vibration tests require artificial excitation induced by heavy shakers and a complete halt to the normal operations of the bridge during the test campaign. In contrast, ambient vibration tests do not require heavy equipment to excite the structure and allow normal operation of the bridge to continue while the test campaign is in session. For these reasons, ambient vibration tests were the preferred option for modal testing on the Boomslang, however, relatively longer periods of data collection are required and the signal levels could be considerably low during these tests (Jaishi & Ren, 2005).

4.3.1 Equipment

In November 2017, a walk-over test was conducted using a vibration application installed on the author's mobile phone. Following this, a full ambient vibration test was conducted on the footbridge. The equipment used for the test included 12 uni-directional piezoelectric accelerometers, signal cables, a 24-channel Data Physics acquisition system and an analogue-to-digital convertor. Accelerometers convert the ambient vibration response into an electrical signal. Signal cables are used to transmit these signals to the analogue-to-digital convertor where the signal is converted into digital data and stored on the hard disk of the data acquisition computer. An exhaustive list of all the equipment used for the ambient vibration test is shown in **Table 4-2**.

Table 4-2: Equipment list for AVT

Name of item	Quantity
Laptop (with charger)	1
Signal Analyser (SA)	1
Analogue to Digital Converter (ADC)	1
Cable Drums	4
Cables (for accelerometers)	12
Accelerometers	12
Coverter Cables	12
Balance Plates	4
Power Cables (SA + ADC)	2
Extention Lead Cables	2
Generator	1
Level	4
Total number of items	**56**

4.3.2 Layout

Due to the perculiar geometry of the bridge, a dense measurement grid guided by the mode shapes obtained from the FE model was defined. This improved the ability to capture most relevant modes of vibration. Accelerometers were placed at 6m intervals beginning on the left edge of the bridge. Subsequent set-ups involved shifting the accelerometers to the right edge, then back to the left edge but ahead of the previous measurement positions. The process continued in a zig-zag manner until the full length of the footbridge was covered.

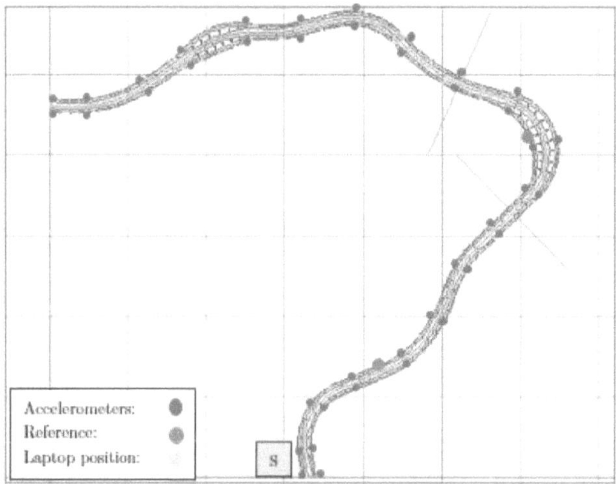

Figure 4-4: Ambient testing layout

Two reference locations with four (two veritcal and two radial) reference accelerometers were selected based on the mode shapes obtained from the preliminary finite element model. The reference locations are denoted by the green dots in **Figure 4-4**. Each set-up yielded a total of 12 sets of data, eight (8) sets from the roving accelerometers and four (4) sets from the reference accelerometers. Once the data is collected from one set-up, the roving accelerometers were shifted to the next location on the bridge while the reference accelerometers remained stationary. It is also important to note that measurements in the vertical and radial direction were conducted separately at the same locations. Finally, the signal analyser, digital converter and the laptop were set up at position *S* shown in **Figure 4-4.**

A 24-channel Data Physics SignalCalc digital spectrum analyser was used to process the response signals. Although the bridge is considered to be relatively lively in the lateral direction, an appropriate frequency resolution is still required to accurately evaluate the modal parameters of the bridge. Therefore, a suggested, yet limited acquisition time of 600s was chosen. This was due to the location and popularity of the structure. Given that Zivanovi, Pavic & Reynolds (2006) obtained satisfactory results using the same acquisition time, it was deemed appropriate to use the same duration for this particular test. The resulting frequency resolution was 0.002Hz (obtained using **Equation 4-1**). The recorded data was sampled at 38.4Hz corresponding to a time-step of 0.026s. A tabulated summary of these parameters is shown in **Table 4-3**.

$$\Delta f = \frac{1}{\Delta T}$$	**Equation 4-1**

Table 4-3: Data acquisition parameters for ambient vibration survey in the vertical and lateral direction

Parameter	Value
Acquisition time (s)	600
Frequency resolution (Hz)	0.002
Sampling frequency (Hz)	38.4
Time step (s)	0.0260417

4.4 Modal Analysis

The modal analysis of the vibration data was performed using ME Scope to obtain the operational modal parameters of the Boomslang. Two main groups of output-only modal identification methods exist: non-parametric (frequency domain based methods) and parametric methods (time domain based methods) (Rainieri & Fabbrocino, 2014). Two methods were implemented in this study: Peak picking (PP) method – a frequency domain technique and stochastic subspace identification (SSI) method- a time domain technique. A detailed discussion about these techniques can be found in Peeters (2000).

4.5 Pedestrian interaction tests

Full-scale pedestrian tests performed by Dallard, Flint, et al., (2001), Brownjohn et al., (2004) and Caetano et al., (2010) revealed that distinct pedestrian loading cases (i.e. groups of pedestrians or continuous stream of pedestrians) have varying effects on the vibration intensity of the footbridge. The tests planned for the Boomslang bridge seeked to investigate the varying effects of groups of pedestrians and a continuous stream of pedestrians on the bridge. Ultimately, the vibration data collected from the continuous stream of pedestrians' test will be used to evaluate the critical number of pedestrians required to trigger lock-in as per the Setra and HiVoSS guidelines.

4.5.1 Layout

The accelerometer locations for the pedestrian tests are denoted as red dots as shown in **Figure 4-5**. The location of the set-up station for the data acquisition system is denoted as *S*. All accelerometer plate locations have three accelorometers each (one vertical, one horizontal and one longitudinal). The channel number or accelerometer number and the direction of the accelerometer are shown in **Table 4-4**.

Table 4-4: Accelerometer orientation

Direction	Accelerometer #
Vertical	1,4,7,10
Radial	2,5,8,11
Longitudinal	3,6,9,12

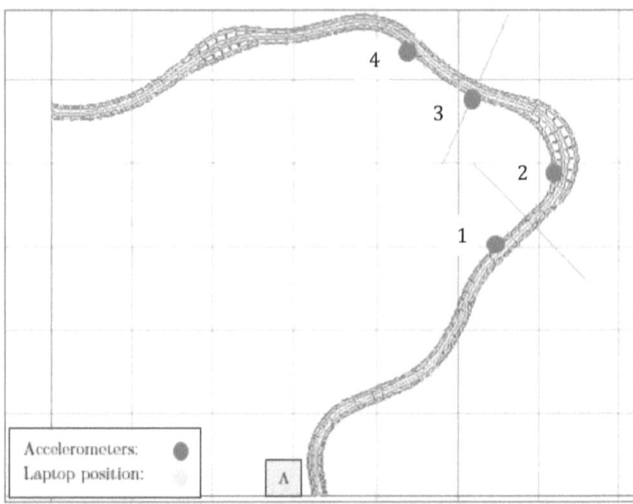

Figure 4-5: Layout of accelerometer plates for pedestrian interaction tests. Each plate hosts a set of 3 accelerometers (i.e Plate 1- Acc 1,2,3; Plate 2- Acc 4,5,6; Plate 3- Acc 7,8,9; Plate 4- Acc 10,11,12)

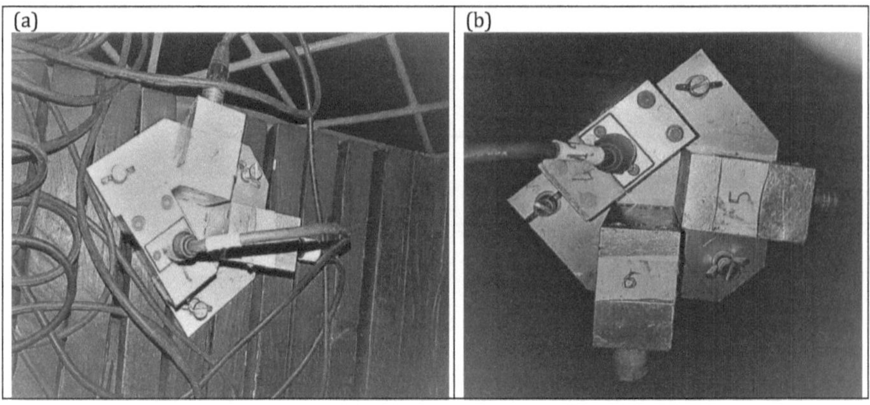

Chapter 4 – Methodology

Itumeleng Ragoleka

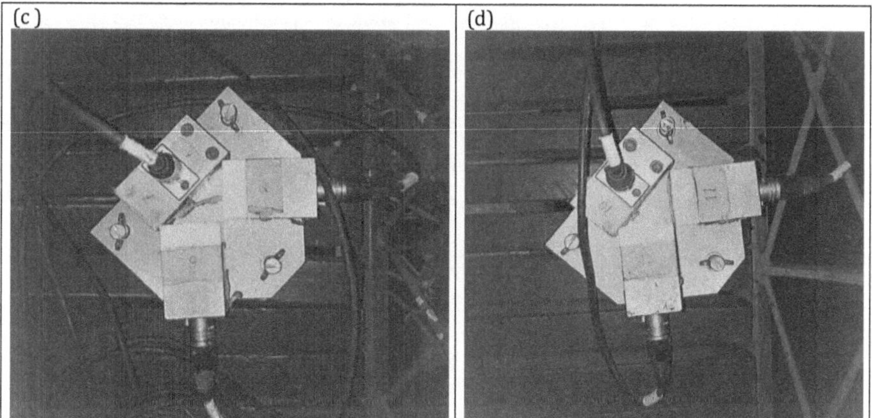

Figure 4-6: (a) – Accelerometer plate 1 hosting Acc 1 (vertical), Acc 2 (lateral) and Acc 3 (longitudinal); (b) – Accelerometer plate 2 hosting Acc 4 (vertical), Acc 5 (lateral) and Acc 6 (longitudinal); (c) – Accelerometer plate 3 hosting Acc 7 (vértical), Acc 8 (lateral) and Acc 9 (longitudinal); (d) – Accelerometer plate 4 hosting Acc 10 (vertical), Acc 11 (lateral) and Acc 12 (longitudinal).

4.5.2 Initial planned test procedure

Full-scale pedestrian tests were conducted on the Boomslang Canopy walkway to determine the critical number of pedestrians required to trigger lateral instability on the bridge. 50 students were recruited for the test campaign. The following tests were conducted:

Test 1: A single pedestrian would be instructed to walk from the start to the end of the bridge along the center line of the bridge.

Test 2: 5 groups of 10 students each would be instructed to walk over the bridge with a 10s interval between each group. All groups of students would be encouraged to walk normaly (i.e at their own comfortable pace).

Test 3: 5 groups of 10 students each would be instructued to walk over the bridge with a 10s interval between each group. All groups of students would be encouraged to walk in synchrony with each other. A metronome would be used to synchronize the students at a pacing rate of 1.8Hz.

Test 4: The students would be instructed to walk as a continuous stream of pedestrians. The students would be encouraged to walk normally (i.e at their own comfortable pace).

Test 5: The students would be encouraged to walk as a continuous stream of pedestrians. The students would be encouraged to walk in synchrony with each other. A metronome would be used to synchronize the students at a pacing rate of 1.8Hz.

4.5.3 Actual test procedure

Due to unforeseen circumstances on site, and the fact that only 39 students attended the investigation session, the planned test procedure was altered. The tests conducted were as follows:

Test 1: 3 groups of 10 students each and 1 group of 9 students were instructed to walk over the bridge with 10s intervals between each group. Three circuits were completed. All groups were encouraged to walk normally (i.e to walk at their own comfortable pace).

Test 2: The students were encouraged to walk as a continuous stream of pedestrians. Two circuits were completed. The students were encouraged to walk normally (i.e at their own comfortable pace).

Figure 4-7: (a) – Test 1; and (b) – Test 2

4.6 Comfort level evaluation according to design guidelines

The peak accelerations recorded during Test 1 and Test 2 were evaluated using Matlab 2016a and compared to the Setra and HiVoSS comfort limits. This was done to evaluate the comfort levels of the bridge according to the design guidelines when subjected to pedestrian loading from the recruited group of students.

4.7 Evaluating the critical number of pedestrians

The critical number of pedestrians will be determined using the Arup model and the vibration comfort limit method established by the design guidelines. Operational modal frequency and damping, as well as modal mass obtained from the FE model will be used when determining the critical number of pedestrians using the Arup model.

4.8 Synchrosqueezed Wavelet Transform

Currently, the frequency content of vibration data is presented in a stagnant form by the Fourier spectrum. Considering that one objective of this book is to comment, with evidence, on whether SLE is the initiating mechanism for excessive lateral vibrations on a footbridge, the synchrosqueezed wavelet transform will be used to monitor the dynamic nature of the fundamental lateral frequency of the footbridge. This result will be used in conjuction with the recorded vibration data to comment on the assumption that SLE is the initiating mechanism for excessive lateral vibrations.

Chapter 5

5 Results

5.1 FE model results

The Sofistik software was used to generate a finite element model of the Boomslang. The undeformed structure is shown in **Figure 5-1**. An assumed target damping ratio of 0.4% (Setra, 2006) was employed in the dynamic analysis of the model. The model was used to calculate the first 10 eigenvalues in an attempt to obtain natural frequencies between 0Hz-10Hz. This range contains the first 5 harmonics of the walking frequency of the pedestrian (Ingólfsson, Georgakis & Jönsson, 2012). The results illustrated in **Figure 5-2** show 8 modes of vibration of the bridge which are below 10Hz. The first 6 modes appear in the range of 0Hz to 5Hz. **Figure 5-3** presents a detailed list of the modal parameters of the structure.

The model shows the first 2 modes as pure lateral modes at f_1=0.898Hz and f_2=2.593Hz. The first mode is anti-symmetric while the second mode is symmetric. The first mode occurs within the detrimental frequency range regarding risk of resonance as established by the Setra and HiVoss guidelines. In contrast, the second mode occurs in range 4 which poses negligible risk to resonance (Setra, 2006).

Mode 3 to mode 8 show complex mode shapes involving a combination of vertical, lateral and torsional modes. No pure vertical modes were found in this frequency range. The complexity of these mode shapes make it difficult to categorize them as either symmetric or anti-symmetric modes. Mode 3 and mode 4, similar to mode 5 and mode 6, are closely spaced modes within a minute frequency range (i.e. 3.306Hz and 3.478Hz; 4.099Hz and 4.413Hz). Mode 3 to mode 6 occur in range 3 on the risk of resonance scale established by the guidelines. Furthermore, mode 7 and mode 8 with frequencies 5.834Hz and 8.808Hz occur in range 4 of the risk of resonance scale.

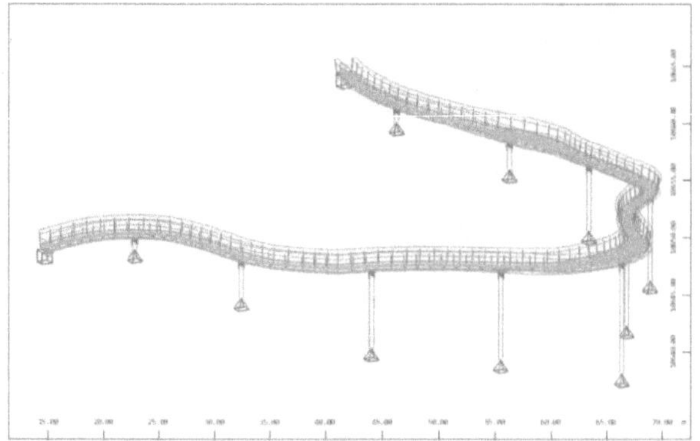

Figure 5-1: Undeformed structure

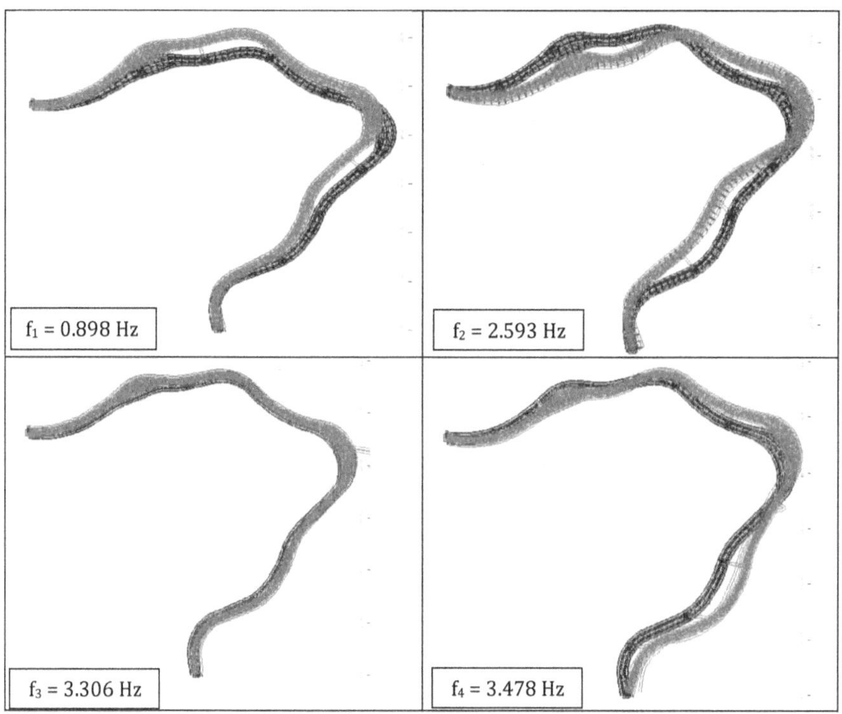

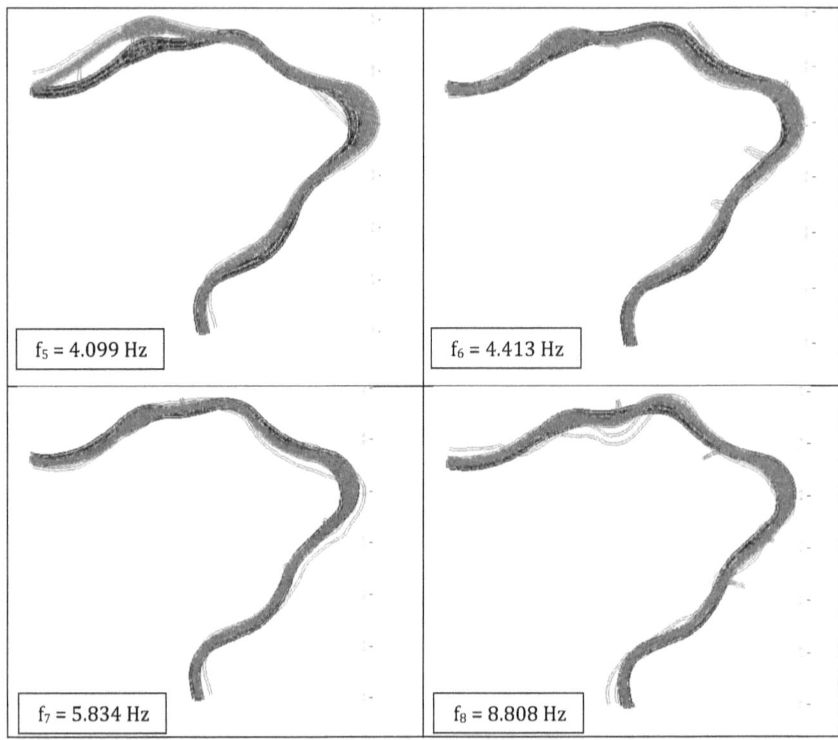

Figure 5-2: First eight numerical vibration modes below 10Hz of the Boomslang Canopy walkway.

```
Control Information
Using Lanczos Method
Iteration vectors                    10
```

Eigenvalues

No.	LC	λ [rad2/sec2]	error [-]	ω [rad/sec]	f [Hz]	T [sec]	ξ [%]	Meff X[%]	Meff Y[%]	Meff Z[%]	participation X[%]	participation Y[%]	participation Z[%]
1	1	3.1833E+01	0.0E+00	5.642	0.898	1.114	0.080	21.8	55.7	0.0	21.8	55.8	0.0
2	2	2.6538E+02	3.1E-05	16.290	2.593	0.386	0.118	2.0	1.5	0.0	2.0	1.5	0.0
3	3	4.3140E+02	1.8E-02	20.770	3.306	0.303	0.485	2.5	0.7	0.0	2.5	0.7	0.0
4	4	4.7759E+02	3.4E-02	21.854	3.478	0.288	1.297	28.8	5.3	0.0	28.9	5.3	0.0
5	5	6.6328E+02	1.1E-01	25.754	4.099	0.244	9.782	12.4	15.6	0.0	12.4	15.7	0.0
6	6	7.6880E+02	2.6E-01	27.727	4.413	0.227	52.434	0.5	5.8	0.0	0.5	5.8	0.0
7	7	1.3436E+03	2.7E-01	36.655	5.834	0.171	42.745	1.5	0.0	0.1	1.5	0.0	0.1
8	8	3.0627E+03	3.8E-01	55.342	8.808	0.114	56.221	0.5	0.0	3.3	0.5	0.0	3.3
9	9	1.0073E+04	9.7E-01	100.362	15.973	0.063	205.42	0.4	1.5	4.4	0.4	1.5	4.4
10	10	2.4980E+06	1.3E+01	1580.503	251.545	0.004	2309.1	0.0	0.0	0.1	0.0	0.0	0.1
							Σ(%)¹	70.4	86.1	7.8	70.5	86.3	7.9

¹ Total effective mass / participation factors of activated masses in X-, Y- and Z-direction.

No.	eigenmode number	f	eigenfrequency
LC	load case	T	eigenperiod
λ	eigenvalue	ξ	modal damping ratio
error	relative eigenvalue error	Meff	effective modal mass in X-, Y- and Z-direction
ω	circular eigenfrequency	participation	participation factors in X-, Y- and Z-direction

Figure 5-3: Numerical estimation of the modal parameters of the Boomslang

5.2   Ambient vibration results

Operational vibration mode shapes and modal characteristics are essential for the understanding of the operational behaviour of the bridge (Rent & Zong, 2004). A table of results showing modal properties between 0Hz and 10Hz is shown in **Figure 5-4**. These were obtained using the ME Scope software. The author has screenshot the views of relevant modes in a perspective that illustrates the dominant mode shape more appropriately.

The results show eight pure lateral modes between 0.868Hz and 3.52Hz. Only the first three lateral modes, excluding mode 1 which is a stationary mode, are illustrated in **Figure 5-4**. It is difficult to classify these mode shapes as either symmetrical or anti-symmetrical due to the multi-curve nature of the bridge. The damping ratios of these modes range from 0.02% to 0.5%. The first lateral mode, which is of particular interest in this study, has a natural frequency of 0.868Hz and an associated damping ratio of 0.101%. The implications of this mode deserve attention since the natural frequency, associated with a low damping ratio, lies in the detrimental range for risk of resonance as established by the Setra and HiVoss guidelines.

The remaining modes with high frequencies are a combination of vertical and torsional modes. These modes range from 4.13Hz to 6.98Hz. Only mode 10, 11 and 12 are illustrated in **Figure 5-4**. The associated damping ratios range from 0.00324% to 0.485%. Distinguishing the level of contribution of each vibration direction in the mode was difficult to achieve, however, the torsional contribution seems to be more dominant than the vertical contribution. This is largely due to the multi-curve nature of the bridge. In fact, no pure vertical modes were observed. Furthermore, it was difficult to classify these modes as either symmetrical or anti-symmetrical modes. According to the Setra and HiVoSS guidelines, this frequency range is safe as it lies beyond the 2nd harmonic walking frequencies of both the lateral and vertical directions.

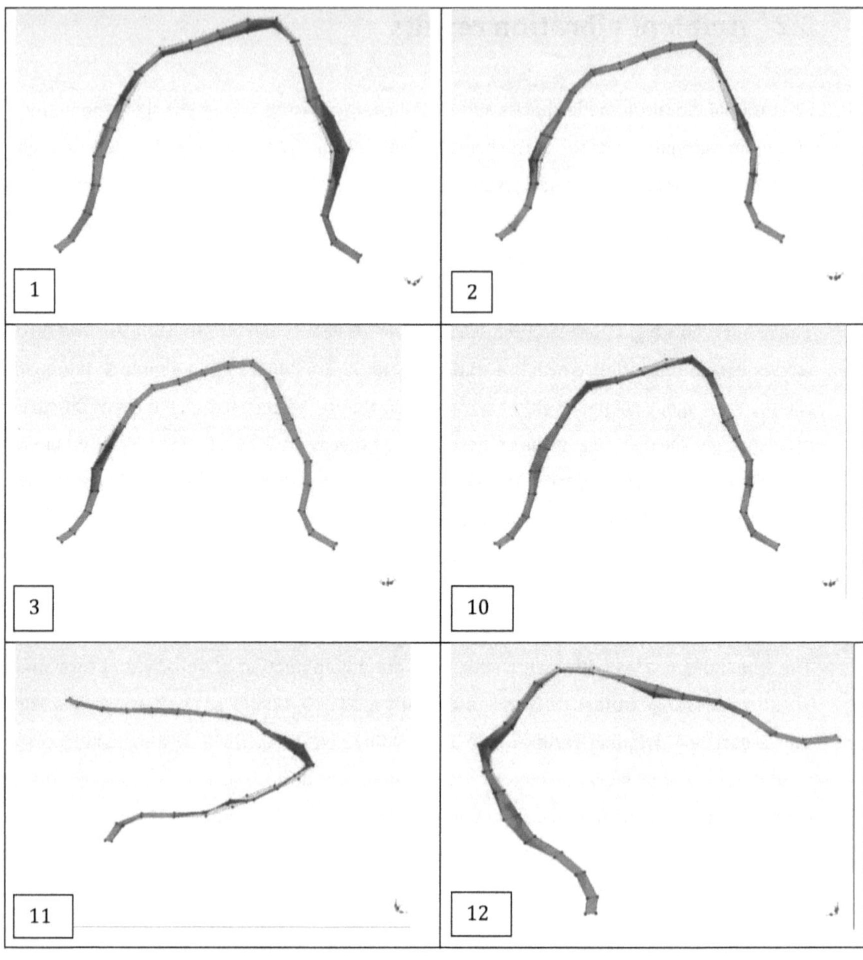

Chapter 5 - Results
MSc book – Itumeleng Ragoleka

Select Mode	Frequency (Hz)	Damping (Hz)	Damping (%)
1	0.0159	0.00115	7.22
2	0.868	0.000873	0.101
3	1.43	0.0077	0.54
4	2.16	0.00544	0.252
5	2.37	0.00475	0.2
6	2.52	0.00288	0.114
7	2.7	0.000613	0.0227
8	3.14	0.00585	0.186
9	3.52	0.00272	0.0773
10	4.13	0.00115	0.0279
11	4.79	0.00115	0.0239
12	5.2	0.000169	0.00324
13	5.43	0.0263	0.485
14	5.65	0.0168	0.298
15	5.99	0.00282	0.0471
16	6.16	0.0022	0.0357
17	6.98	0.00222	0.0318

Figure 5-4: Illustrations of the first 3 lateral modes (mode 2, 3, 4) and the first 3 torsional modes (modes 10, 11, 12). Results table shows all modes between 0Hz and 10Hz.

5.3 Crowd investigation results

The pedestrian interaction investigations were conducted on the 23rd of March 2018 at the SANBI Kirstenbosch National Botanical Gardens. The gardens and the Boomslang Canopy walkway are very popular and thus the investigations could only be conducted after 17:00pm to minimize disturbances to visitors of the bridge. A class of 50 students were initially recruited for the pedestrian interaction investigations, however, only 39 students attended the investigation session. The signed register of students is attached in the **Appendix**.

Considering the core topic of this book, primary attention will be awarded to discussing the lateral vibration results from the pedestrian interaction investigations. For the purpose of completeness, vertical vibration results will also be summarized at the end of this section.

5.3.1   Test 1

The instructions for Test 1 were that the students organize themselves into groups of 10 members each and the last group with only 9 members. All groups were let onto the bridge with 10s intervals between each group. The complete test period was 1200s. This instruction was maintained for two rounds in the first 600s of the experiment. Beyond this moment, the students exited the bridge, returned to the starting point and traversed the bridge for the third time in the remaining 600s of the test. The cluster formation of the groups and the 10s interval between each group for the last 600s of the test was not as accurate as in the first 600s of the test.

Figure 5-5 shows the time domain response from the investigation described above. The groups of students, equating to 0.3p/m², traversed over the bridge at approximately 1.1m/s with the last student exiting the bridge after approximately 170s (i.e. one round). This result assumes that the students are let onto the bridge every 0.5s with a 10s interval between each group of ten students. The response data shows three distinct response growth and response decay sections. Each section lasts for a duration of approximately 300s. As shown in **Figure 5-5**, two sections, excluding channel 8, exceed all comfort limits established by the Setra and HiVoSS guidelines.

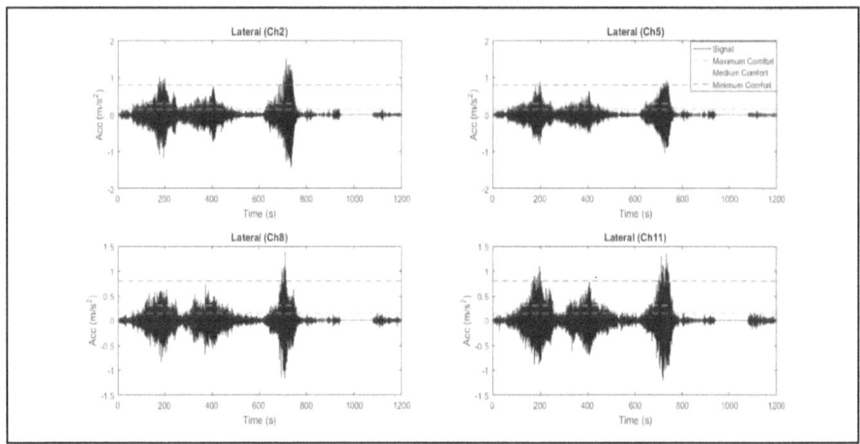

Figure 5-5: Lateral acceleration results per laterally oriented accelerometer for Test 1. Comfort limits established by the Setra and HiVoSS are indicated by the horizontal dashed lines.

Table 5-1 shows the maximum recorded lateral accelerations while **Figure 5-6** illustrates these results graphically. **Figure 5-6** also compares the maximum recorded lateral accelerations to the established comfort limits. All recorded maximum lateral accelerations exceeded all comfort limits. **Figure 5-7** shows the magnitude difference of the maximum recorded accelerations to the relevant comfort limit.

Since all comfort limits were exceeded, the more important result is the magnitude difference between the maximum recorded acceleration and the minimum comfort limit. Accelerometer 2 recorded the highest exceedance to the minimum comfort limit (i.e. 0.71m/s²) while accelerometer 5 recorded the least exceedance to the same comfort limit (i.e. 0.12m/s²).

Table 5-1: Maximum lateral response per accelerometer for Test 1

Test 1			
Lateral response			
Ch	Max acceleration (m/s²)	Max comfort limit (m/s²)	Mean comfort limit (m/s²)	Min comfort limit (m/s²)
2	1.5070	0.15	0.3	0.8
5	0.9247	0.15	0.3	0.8
8	1.3892	0.15	0.3	0.8
11	1.3549	0.15	0.3	0.8

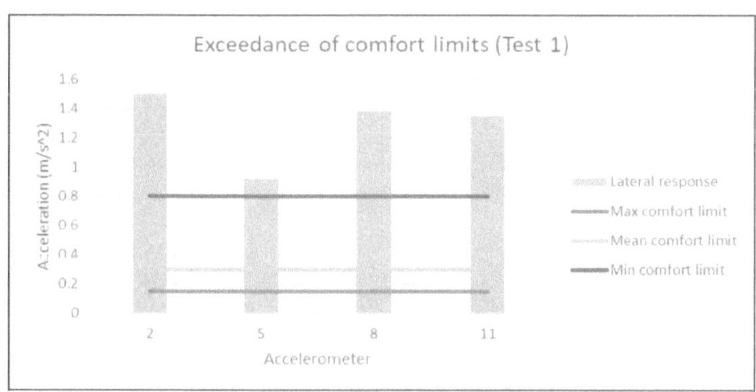

Figure 5-6: Exceedance of comfort limits pertaining to Test 1

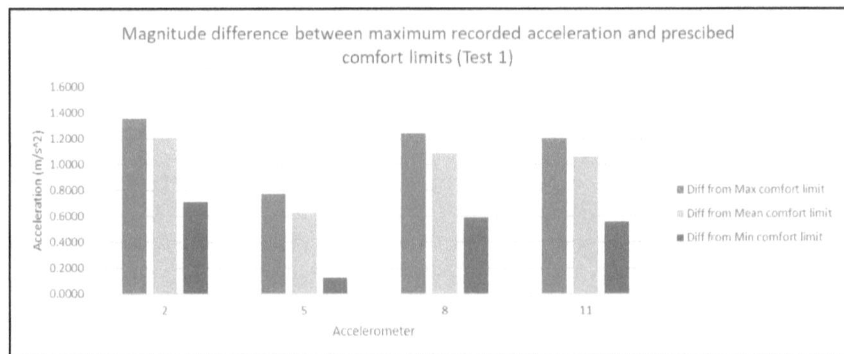

Figure 5-7: Magnitude difference between maximum recorded acceleration per laterally oriented accelerometer and the relevant comfort limit.

Figure 5-8 shows the frequency spectrum of the lateral response data. All four lateral accelerometers show a fundamental frequency of approximately 0.86Hz. This is followed by natural frequencies of 1Hz and 2.5Hz. Beyond 2.5Hz, no other dominant natural frequencies are observed. By scale, it is observed that the 0.86Hz component is dominant at channel 2 with a psd amplitude of 0.18, while being the least at channel 8 with a psd amplitude of 0.03. However, the 1Hz component is more dominant at channel 8 with a psd amplitude of 0.07 and the least at channel 2 with a psd amplitude of 0.01.

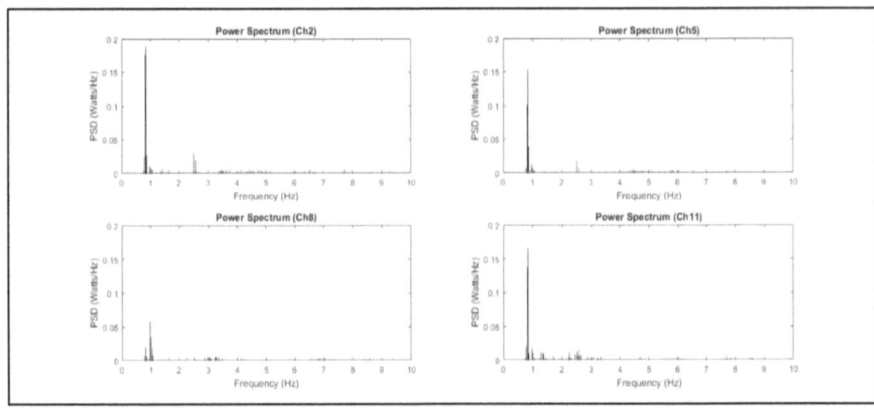

Figure 5-8: Power spectral density plot for Test 1 results

Figure 5-9 below shows the wavelet synchrosqueezed transforms for Test 1 results. Across all accelerometers, a dominant frequency of approximately 0.86Hz is observed at four distinct intervals; i.e. between 2 minutes (120s) and 4 minutes (240s), 6 minutes (360s) and 8 minutes (480s), 9 minutes (560s) and 10.5 minutes (630s), and 11 minutes (660s) and 13 minutes (780s). These intervals correspond with the response increase and response decay sections observed in **Figure 5-5**. The plot referring to channel 8 shows another distinct frequency at 1Hz during the same time intervals. According to the intensity bar, the intensity of these plots ranges between 0.1 and 0.9. The highest intensity is observed during the interval of 11 minutes to 13 minutes. The varying level of intensity of the wavelet ridge may suggest that the relevant frequency was not significantly engaged during that particular interval. Natural frequencies above 1Hz seem to be less concentrated and thus suggest that those frequencies were not dominant in the investigation.

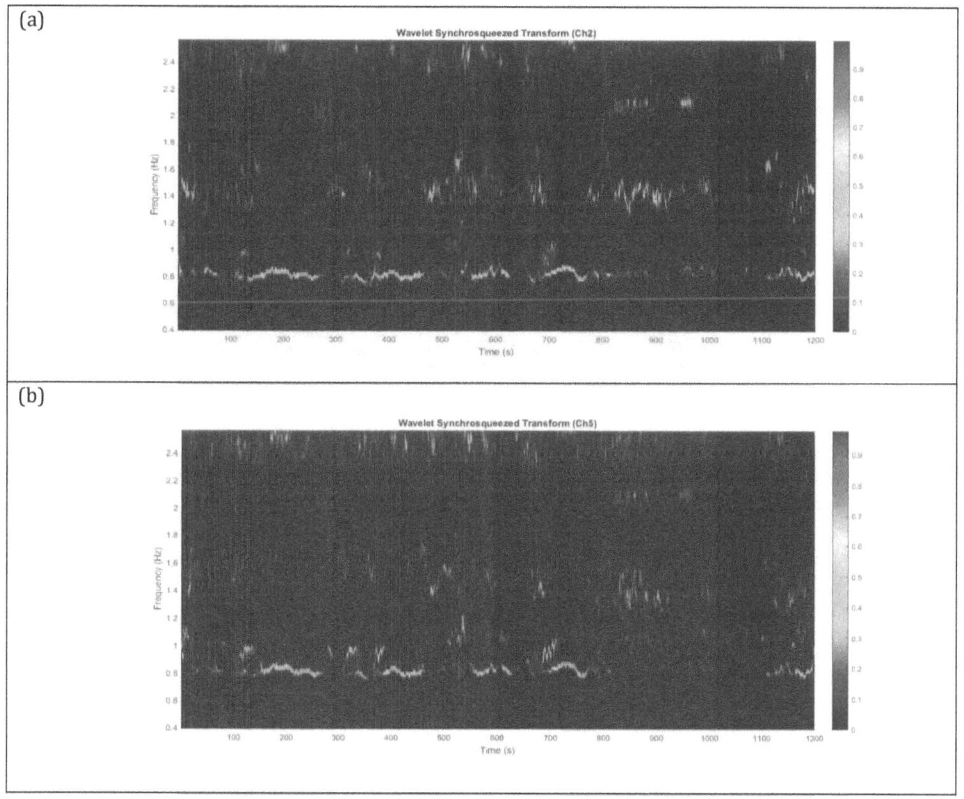

Chapter 5 - Results
MSc book – Itumeleng Ragoleka

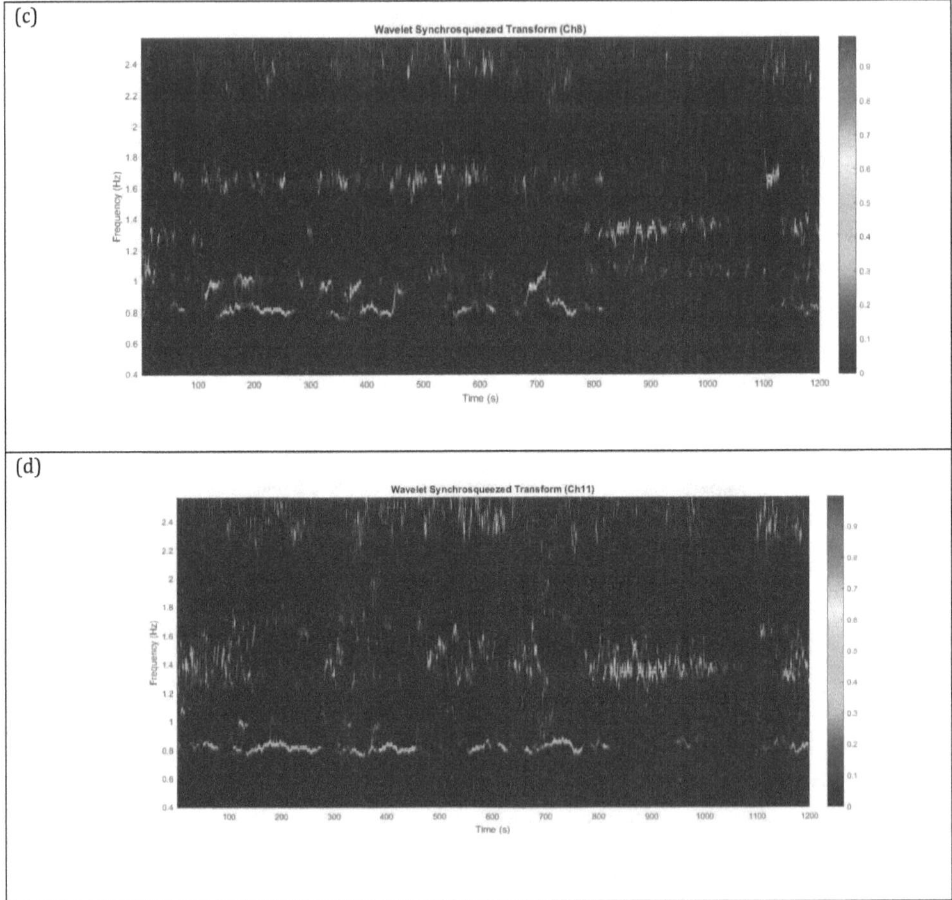

Figure 5-9: Wavelet synchrosqueezed transform for Test 1. (a) – channel 2; (b) – channel 5; (c) – channel 8; and (d) – channel 11.

Figure 5-10 shows the decomposed (first mode and second mode) signals recorded during Test 1-channel 8. Both vibration responses show peaks below the maximum comfort limit in the first 600s of the test. However, both responses exceed the maximum comfort limit in the latter 600s of the test.

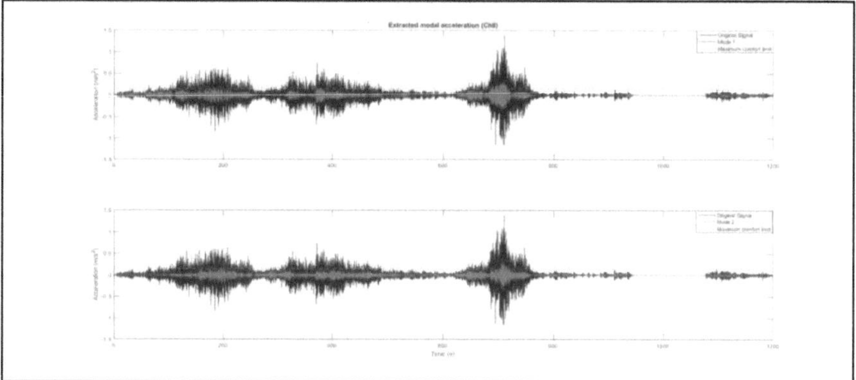

Figure 5-10: Measured 1st and 2nd mode decomposed signal from accelerometer 8 during Test 1

Due to unforeseen circumstances and time limitations concerning organised transport for the volunteering students, the subsequent tests were shortened to 600s of recording. Therefore, for the purposes of appropriate comparison hereafter, the results recorded during the first 600s of Test 1 will be considered.

Table 5-2 shows the maximum lateral vibration levels recorded during the first 600s of Test 1. Similarly, **Figure 5-11** shows graphically the results presented in **Table 5-2**. **Figure 5-11** shows that all accelerometers recorded maximum vibration levels exceeding the maximum and mean comfort limits. Maximum recorded vibration levels pertaining to accelerometer 2,5 and 11 also exceeded the minimum comfort limit while accelerometer 8 was below this limit.

Figure 5-12 shows the magnitude difference between the maximum recorded lateral acceleration and the relevant comfort limit. Since accelerometer 2,5 and 11 exceeded all comfort limits, the important result regarding these is the magnitude difference relative to the minimum comfort limit. The highest exceedance was recorded by accelerometer 11 of 0.3m/s² while the least exceedance was recorded by accelerometer 5 of 0.09m/s². Accelerometer 8 exceeded the mean comfort limit by 0.44m/s².

Table 5-2: Maximum lateral response per accelerometer for Test 1- first 600s

Test 1 (first 600s)			
Lateral response			
Ch	Max acceleration (m/s^2)	Max Limit (m/s^2)	Mean Limit (m/s^2)	Min Limit (m/s^2)
2	1.0182	0.15	0.3	0.8
5	0.8922	0.15	0.3	0.8
8	0.7444	0.15	0.3	0.8
11	1.1003	0.15	0.3	0.8

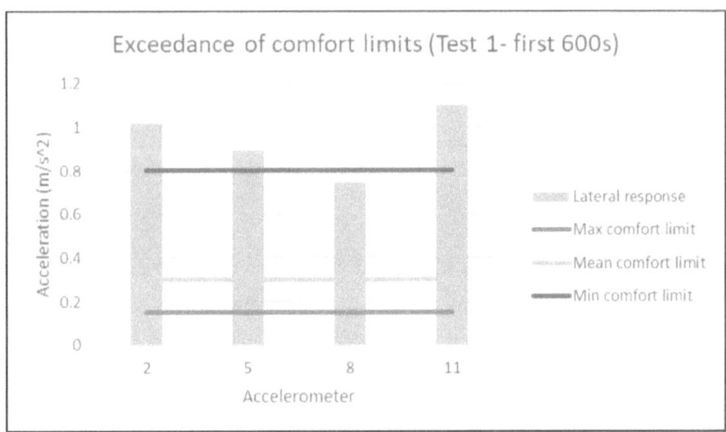

Figure 5-11: Comparison of peak recorded accelerations and established comfort limits for Test 1- first 600s

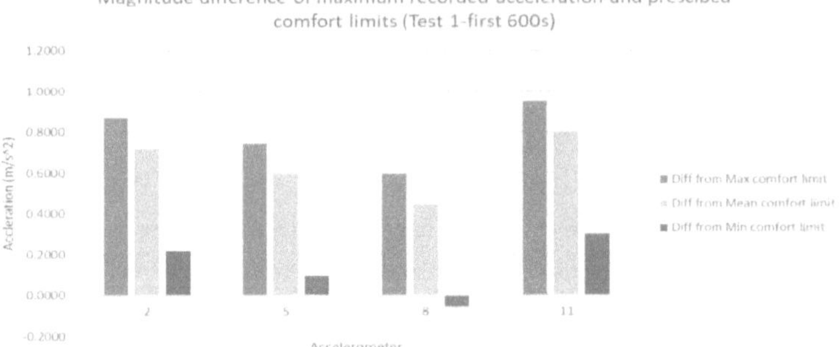

Figure 5-12: Magnitude difference between maximum recorded lateral acceleration during the first 600s of Test 1 and the relevant comfort limit

Figure 5-8, **Figure 5-9** and **Figure 5-10** can also be used to evaluate the results recorded in the first 600s of Test 1. The comments made above for these plots are consistent with the events recorded during the first 600s of Test 1.

5.3.2   Test 2

Test 2 involved continuous random walking of all the students on the footbridge. Two rounds of walking were completed in 600s. **Figure 5-13** shows the time domain lateral response results for Test 2. A single student walking at approximately 1.1m/s crosses the bridge in 115s. A delayed period of approximately 20s is assumed between the first and the last student considering that 39 students were let onto the bridge at a rate of 0.5s per student. Therefore, the class equating to $0.3p/m^2$ is assumed to cross the bridge in approximately 135s.

The response data illustrated in **Figure 5-13** shows two distinct sections of response growth and response decay. Each section extends for a period of approximately 150s each. Accelerometer 2 and accelerometer 5 recorded maximum vibration levels exceeding the maximum and mean vibration comfort limits while accelerometer 8 and accelerometer 11 recorded vibration levels exceeding all vibration comfort limits.

Table 5-3 shows the maximum lateral accelerations recorded by the laterally oriented accelerometers. **Figure 5-14** illustrates graphically the results in **Table 5-3** and compares the maximum recorded response values to the prescribed comfort limits. According to **Figure 5-14**, confirming **Figure 5-13**, accelerometer 2 and accelerometer 5 recorded maximum vibration levels exceeding the mean comfort limit, while accelerometer 8 and accelerometer 11 recorded vibration levels exceeding the minimum comfort limit.

Figure 5-15 shows the magnitude difference between the maximum recorded lateral acceleration and the relevant comfort limit. Concerning accelerometer 2 and 5, the important result is the exceedance value to the mean comfort limit which was $0.45m/s^2$ and $0.47m/s^2$. Regarding accelerometer 8 and 11, the important result is the exceedance value to the minimum comfort limit which was $0.55m/s^2$ and $0.27m/s^2$.

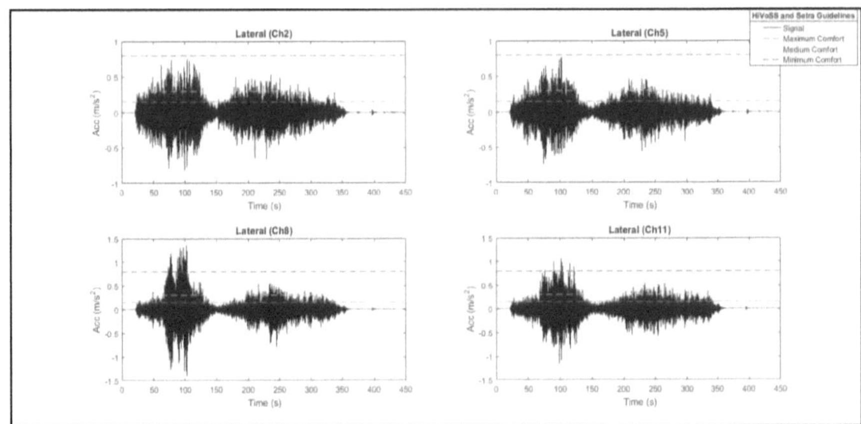

Figure 5-13: Lateral acceleration results per laterally oriented accelerometer for Test 2. Comfort limits are indicated by the horizontal dashed lines

Table 5-3: Maximum lateral acceleration per accelerometer for Test 2

	Test 2			
	Lateral response			
Ch	Max acceleration (m/s^2)	Max comfort limit (m/s^2)	Mean comfort limit (m/s^2)	Min comfort limit (m/s^2)
2	0.7451	0.15	0.3	0.8
5	0.7679	0.15	0.3	0.8
8	1.3501	0.15	0.3	0.8
11	1.0716	0.15	0.3	0.8

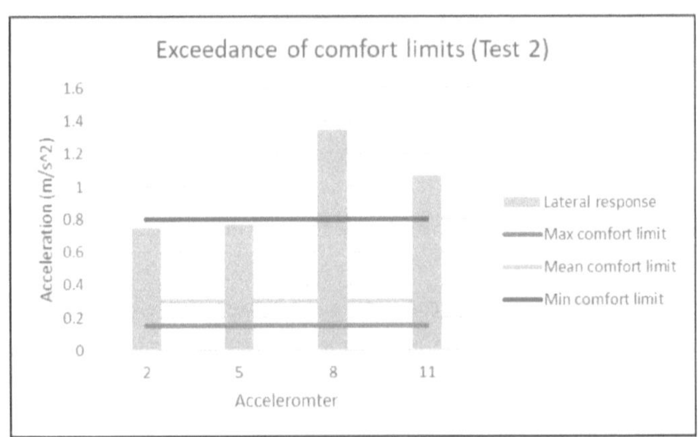

Figure 5-14: Comparison of peak recorded accelerations and established comfort limits for Test 2

Chapter 5 - Results
MSc book – Itumeleng Ragoleka

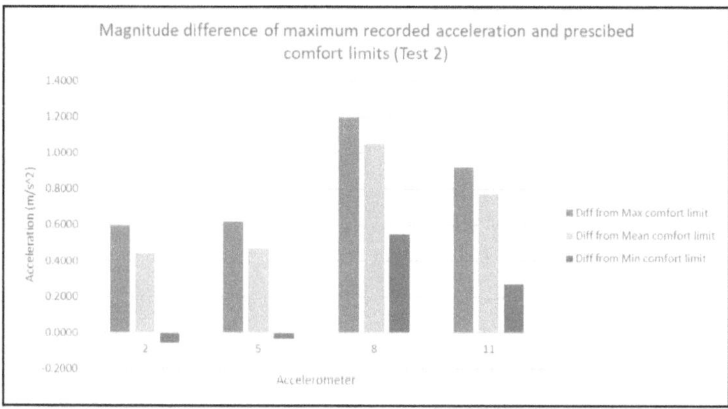

Figure 5-15: Magnitude difference between maximum recorded lateral acceleration for Test 2 and the relevant comfort limit

Power spectral density plots:

Figure 5-16 shows the power spectral density plots of the results pertaining to Test 2. It is observed that there are two dominant fundamental frequencies of approximately 0.80Hz and 1Hz. Other dominant frequencies appearing below 5Hz are 1.3Hz, 2Hz, 3Hz, 3.2Hz, 3.6Hz and 4.5Hz. By scaling, the 1Hz component is observed to be dominant in channel 8 with a psd amplitude of 0.13 and being the least dominant in channel 11 with a psd amplitude of 0.025. On the other hand, the 0.80Hz component is dominant in channel 11 with a psd amplitude of 0.1, while being negligible in channel 8. Furthermore, a dominant 3rd harmonic component is observed at 3Hz in channel 8 with a psd amplitude of 0.08. The same component is observed in channel 11 with a psd amplitude of 0.025, but negligible in channel 2 or channel 5.

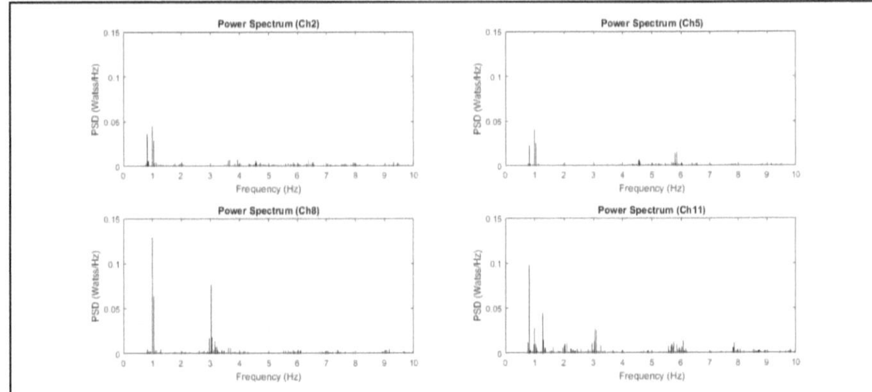

Figure 5-16: Power spectral density plot of Test 2 results

Figure 5-17 below shows the wavelet synchrosqueezed transforms for Test 2 results. Across all accelerometers, two dominant frequencies of approximately 0.80Hz and 1Hz are observed between 1 minute (60s) and 2.5 minutes (150s). The same frequencies reappear with less intensity between 3 minutes (180s) and 6 minutes (360s). According to the intensity bar, the range of intensity in these plots was between 0.1 and 0.9. The highest intensity is observed, in part, for all accelerometers between 1 minute (60s) and 2 minutes (120s). Results pertaining to higher frequencies are highly distorted and do not convey meaningful information about the precise frequencies which are engaged in this test. It is noted that the wavelet ridge experiences a sudden "dip and rise" at approximately 70s for all accelerometers. Again, the discontinuity of the wavelet ridge is consistent with the sections of response increase and response decay observed in **Figure 5-13**.

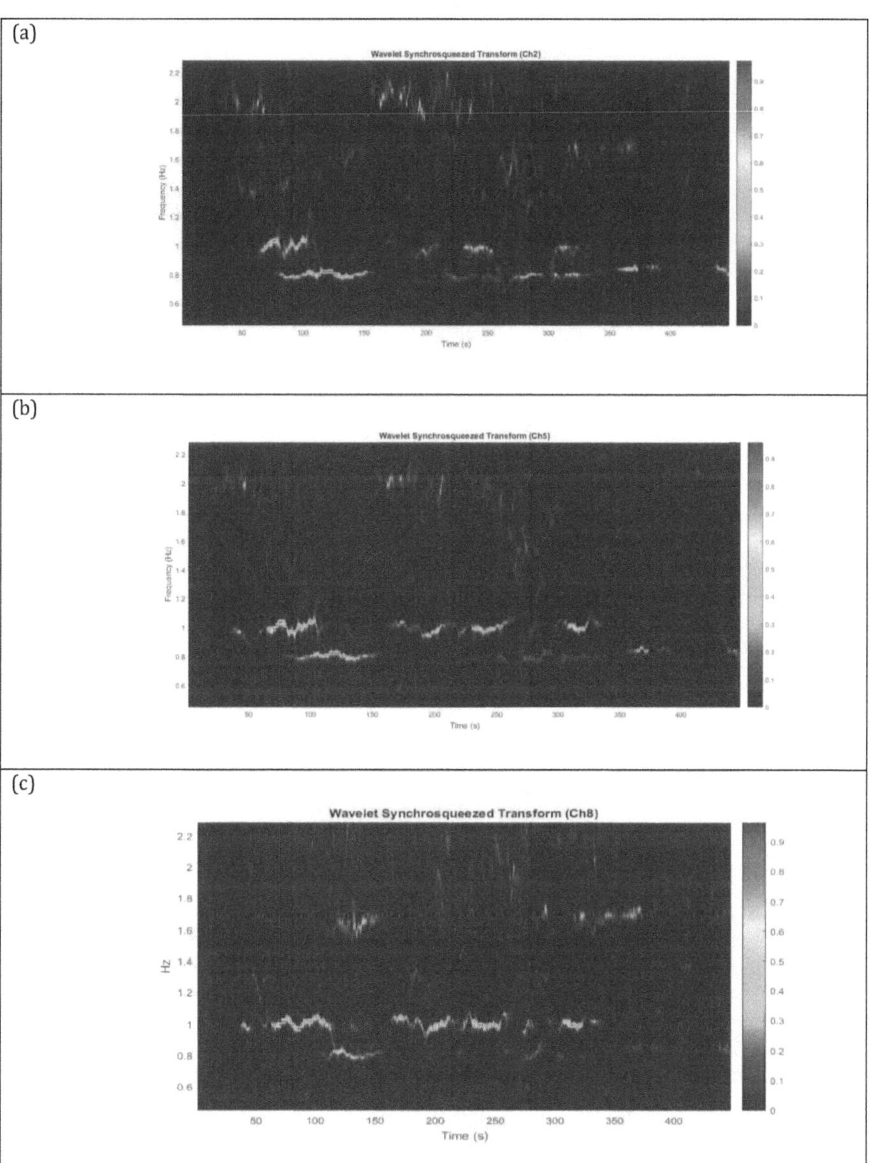

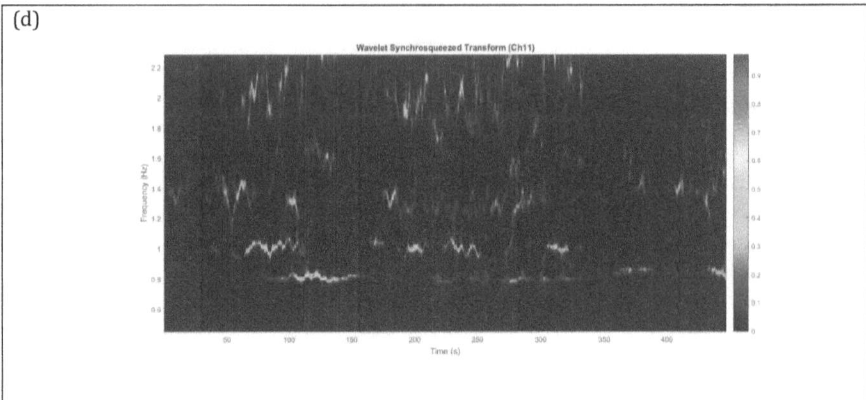

Figure 5-17: Wavelet synchrosqueezed transform for Test 2

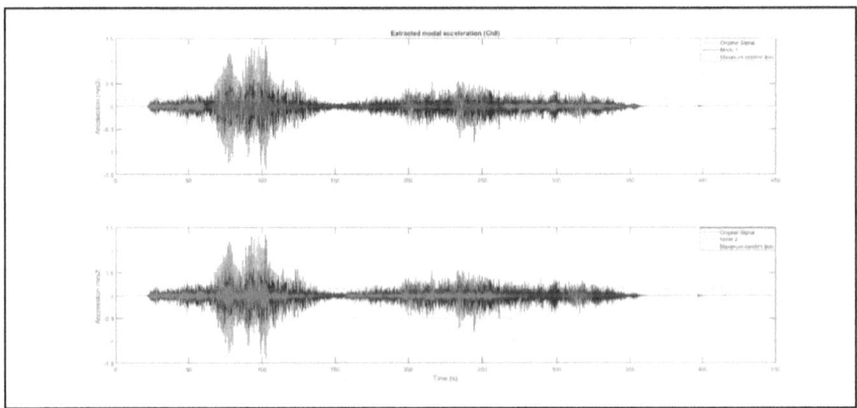

Figure 5-18: Measured 1st and 2nd mode decomposed signal from accelerometer 8 of Test 2

Figure 5-18 above shows the mode decomposed (first mode and second mode) signals recorded by channel 8 during Test 2. The results show that the peak accelerations of both modes exceeded the maximum comfort limit between 60s and 120s.

5.3.3   Critical number of people

ARUP Model:

The Arup model was developed by Dallard, Flint, et al., (2001) with the premise that pedestrians walking on a bridge induce a lateral force which acts as a source of negative damping to the bridge's lateral motion (Dallard, Flint, et al., 2001). The critical condition of this model is defined as the number of pedestrians beyond which the cumulative negative damping force becomes higher than the inherent damping force of the bridge. This model was used to evaluate the critical number of pedestrians required to trigger lock-in on the Boomslang bridge. The required inputs in the model are the fundamental lateral frequency of the bridge, the modal mass and the modal damping ratio. These inputs are obtained from the numerical model and the measured data. A summary of the inputs is provided below.

The assumptions embedded in the Arup model are that the pedestrians are walking at the same lateral frequency as the fundamental frequency of the structure, and that the mode shape of the first mode is sinusoidal. The evaluation of the critical number of pedestrians is presented in **Table 5-4** as 13 pedestrians. This means that 13 pedestrians are required to walk in synchrony with the bridge lateral motion at 0.868Hz to trigger lock-in and cause excessive lateral vibrations on the bridge. This result could either be valid or not given the fact that the first lateral vibration mode of the bridge is not sinusoidal due to the geometry of the structure. However, the parameters presented below are valid; including the lateral force coefficient which has been proven to provide reasonable estimates for any footbridge if the fundamental lateral frequency is between 0.5Hz and 1Hz.

A shortcoming of the Arup model is that it is not capable of determining how long the pedestrians need to walk on the bridge before the lock-in event is initiated. Therefore, although it is determined that 13 pedestrians (equivalent to $0.1p/m^2$) are capable of initiating lock-in, it is not determined how long the pedestrians need to walk on the bridge before the lock-in phenomenon is engaged.

Table 5-4: Modal parameters of the Boomslang

Parameter	Value	Unit
Natural frequency (f_n)	0.868	Hz
Modal mass (m_i)	173802.243	kg
Modal damping ratio (ξ)	0.101	%
Lateral force coefficient (k)	300	Ns/m
Critical number of people (N_p)	13	people

SETRA and HiVoSS Method:

The critical number of pedestrians required to trigger lock-in was determined using the results from Test 2- accelerometer 8. The reason for this choice is that all the reviewed cases by the author of full-scale dynamic testing have determined this value from an investigation involving a continuous stream of pedestrians (Dallard, Flint, et al., 2001; Caetano, Cunha, Moutinho & Magalhás, 2010). Furthermore, accelerometer 8 recorded the highest peak acceleration for Test 2 and will thus provide a conservative result.

Figure 5-19 shows the vibration response of the bridge deck subject to a continuous stream of students during Test 2. The students were all let onto the bridge 10s after the recording of the test was started. The students were loaded onto the bridge for a period of 20s at a rate of 0.5s per student. The students traversed the full length of the bridge for a further 80s. They quickly returned to the starting position and repeated the walk over the bridge. A gap of 50s passed to allow the students to return to the starting point and repeat the walk. A further 80s passed during which the students repeated the walk over the full length of the bridge. From the time spent on the bridge, it seems that the average speed of the students was 1.6m/s.

The response data shows a steady increase in magnitude from 30s to 70s, after which an exponential increase in magnitude is observed showing evidence of magnitudes as high as 1.35m/s². This result is significantly beyond the critical limit of 0.15m/s² suggested by the Setra and HiVoSS guidelines for potential pedestrian lateral synchronization.

The acceleration response from 150s to 400s represents the second walk-over by the students. It is observed that the peak response during this period is lower than that recorded during the first walk-over between 0s and 110s. The gradual increase in dynamic response peaks at approximately

0.5m/s² compared to 1.35m/s² recorded during the first walk-over. This results in a difference in peaks of 0.85m/s². The peak acceleration recorded during the first walk-over is 1.2m/s² higher than the established critical limit for anticipated lateral synchronization of pedestrians while the peak acceleration recorded during the second walk-over is 0.35m/s² beyond the same limit.

Regarding the determination of the critical number of people required to trigger lock-in using the 0.15m/s² limit method suggested by the guidelines, **Figure 5-19** shows that the evidence of exponential growth in vibration response occurs at approximately 70s. However, this occurs once all the pedestrians have loaded the bridge and walked for a reasonable amount of time. The insufficient number of volunteering students disqualifies the possibility to determine an absolute number of people required to trigger lock-in as done by Caetano et al., (2010) and Dallard et al (2001). Alternatively, **Figure 5-19** shows evidence of 39 students being loaded on the bridge during a period of 20s (i.e from 10s to 30s) and walking on the bridge for approximately 40s before the bridge vibration behaviour responds in an exponential manner. As far as this investigation is concerned, the conclusion to the critical number of people required to trigger lock-in on the Boomslang is limited to exactly what was observed: 39 pedestrians walking for approximately 40s. It is important to note that the author observed no evidence of lateral synchronisation amongst the students nor with the vibrating bridge deck during this investigation.

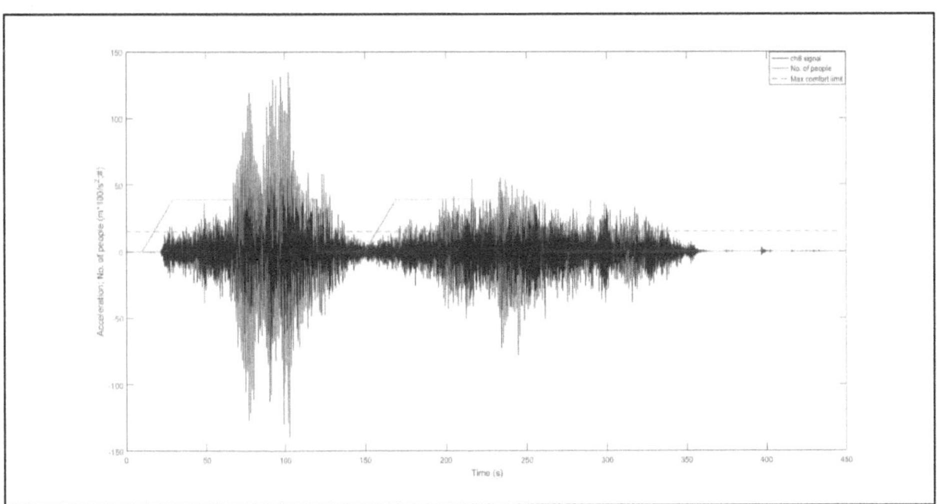

Figure 5-19: Response of bridge deck due to a continuous stream of students.

5.3.4   Vertical response results

Figure 5-20 shows the vertical response data per vertically oriented accelerometer for Test 1. Accelerometers 1,4 and 7 recorded peak accelerations less than the maximum comfort limit in the first 600s of Test 1 as seen in **Table 5-5**. Only accelerometer 10 recorded peak accelerations beyond the maximum comfort limit for the same period. For the last 600s of the test, accelerometer 4 and 10 recorded peak accelerations exceeding the maximum comfort limit but not the mean and minimum comfort limits. **Figure 5-21** (**a** and **b**) shows graphically the peak accelerations recorded by accelerometer 1,4,7 and 10 in the first and last 600s of Test 1. For both the first and last 600s of Test 1, accelerometer 10 recorded the highest peak accelerations of 0.85m/s² and 0.93m/s². From these results, it is clear that the vertical vibrations pose no risk of resonance on the Boomslang.

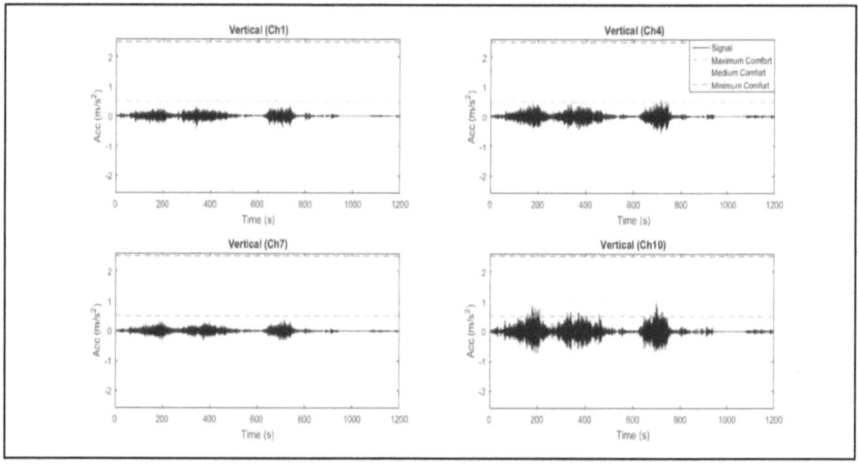

Figure 5-20: Vertical response data for Test 1

Table 5-5: Maximum response levels per accelerometer for Test 1-first 600s

Test 1 (First 600s)			
Vertical response			
Ch	Max acceleration (m/s^2)	Max comfort limit (m/s^2)	Mean comfort limit (m/s^2)	Min comfort limit (m/s^2)
1	0.3132	0.5	1	2.5
4	0.4215	0.5	1	2.5
7	0.3148	0.5	1	2.5
10	0.8534	0.5	1	2.5

Table 5-6: Maximum response levels per accelerometer for Test 1-last 600s

Test 1 (Last 600s)			
Vertical response			
Ch	Max acceleration (m/s^2)	Max comfort limit (m/s^2)	Mean comfort limit (m/s^2)	Min comfort limit (m/s^2)
1	0.3190	0.5	1	2.5
4	0.5846	0.5	1	2.5
7	0.3559	0.5	1	2.5
10	0.9316	0.5	1	2.5

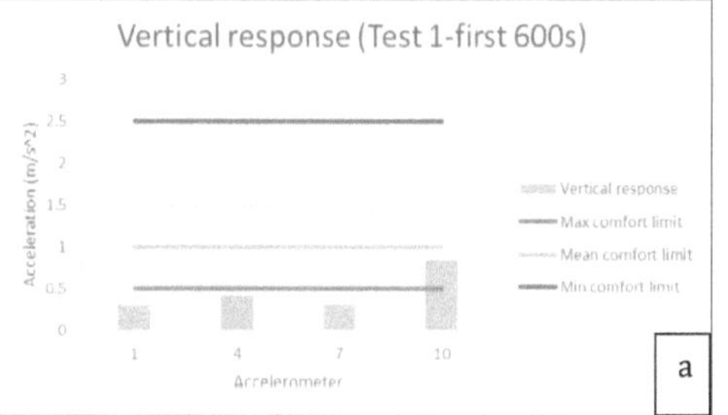

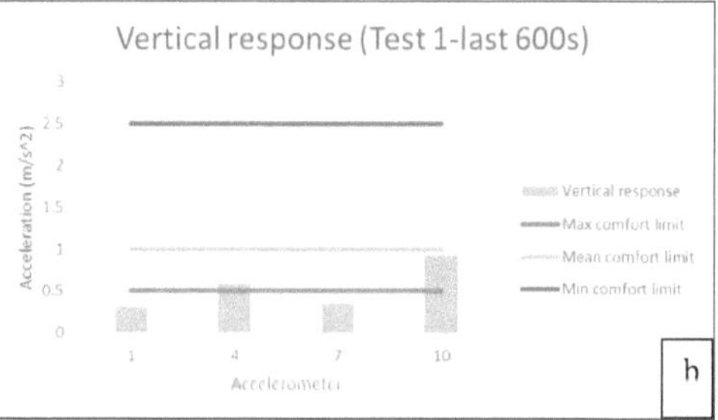

Figure 5-21: Maximum acceleration per accelerometer compared to comfort limits stipulated in design guidelines (a= first 600s; b=last 600s)

Figure 5-22 shows the power spectral density plots of the vertical response data recorded during Test 1. Significant activity is observed beyond 3Hz. Accelerometer 1 shows a minor mode at approximately 2.5Hz, thereafter at 6.9Hz, 8.6Hz and 9.4Hz. Accelerometer 4 shows a minor mode at 4.1Hz, thereafter, at 8.6Hz, 9Hz, 9.4Hz and 9.7Hz. Accelerometer 10 shows a lot more activity than accelerometer 1, 7 and 4. Accelerometer 10 shows modes as low as 0.9Hz. Thereafter, 4Hz, 5.7Hz, 6.9Hz, 7.1Hz, 9Hz and 9.4Hz. In general, channel 10 shows evidence of severe transients experienced during the test.

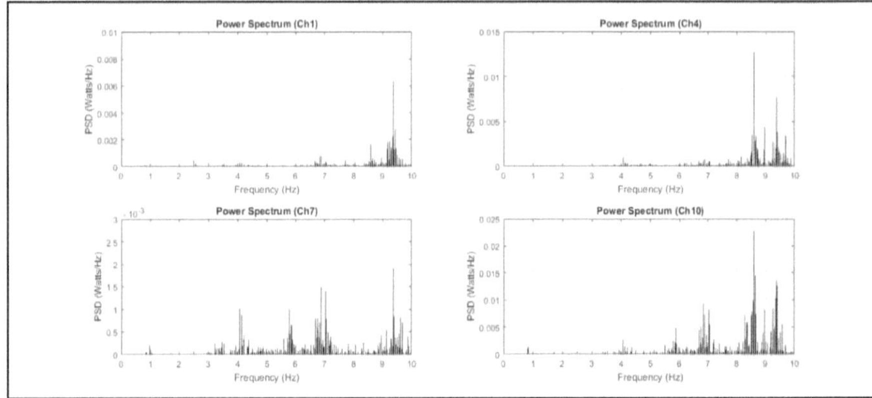

Figure 5-22: Power spectral density of vertical response for Test 1

Figure 5-23 shows the vertical response data per vertically oriented accelerometer for Test 2. Accelerometers 1 and 7 recorded peak accelerations less than the maximum comfort limit as seen in **Table 5-7**. Accelerometer 4 and 10 recorded peak accelerations beyond the maximum comfort limit but did not exceed the mean comfort limit. **Figure 5-24** shows graphically the peak accelerations recorded by accelerometer 1,4,7 and 10. Accelerometer 10 recorded the overall highest peak acceleration of 0.86m/s^2 during Test 2.

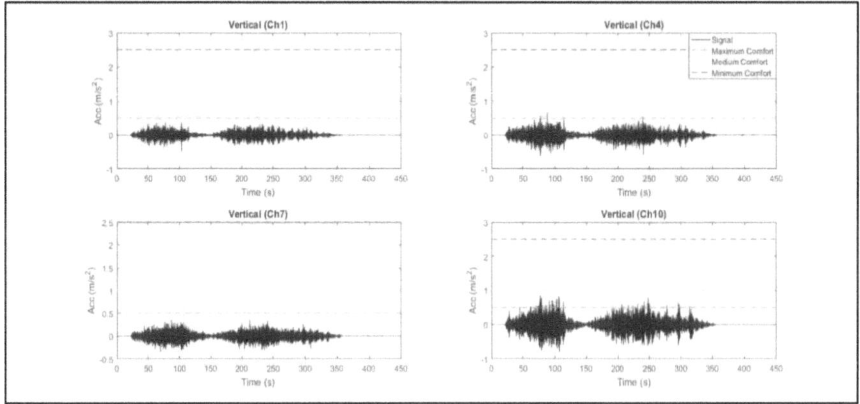

Figure 5-23: Vertical response data for Test 2

Table 5-7: Maximum recorded accelerations per accelerometer for Test 2

Test 2			
Vertical response			
Ch	Max acceleration (m/s^2)	Max comfort limit (m/s^2)	Mean comfort limit (m/s^2)	Min comfort limit (m/s^2)
1	0.3870	0.5	1	2.5
4	0.6777	0.5	1	2.5
7	0.3611	0.5	1	2.5
10	0.8595	0.5	1	2.5

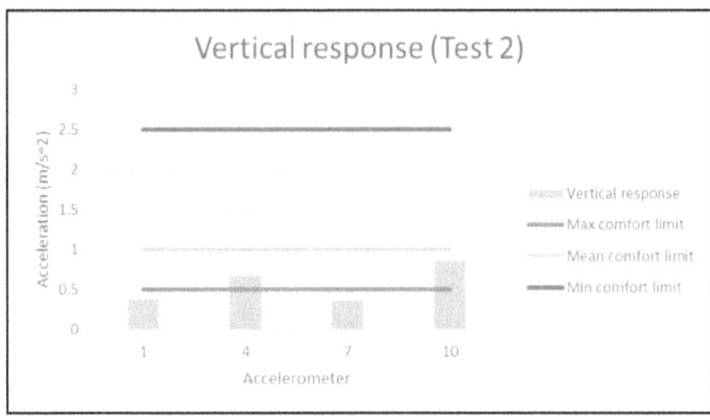

Figure 5-24: Maximum acceleration per accelerometer compared to comfort limits stipulated in design guidelines

Figure 5-25 shows the power spectral density plots of the vertical response data recorded during Test 2. All accelerometers show evidence of a minor mode at 4Hz. Accelerometer 1 shows evidence of a dominant mode at 9.3Hz while accelerometer 4 shows evidence of transients beyond 8Hz. Accelerometer 7 and 10 show evidence of modes at 5.8Hz and 7Hz, however, accelerometer 10, unlike accelerometer 7, shows evidence of transients beyond 8Hz.

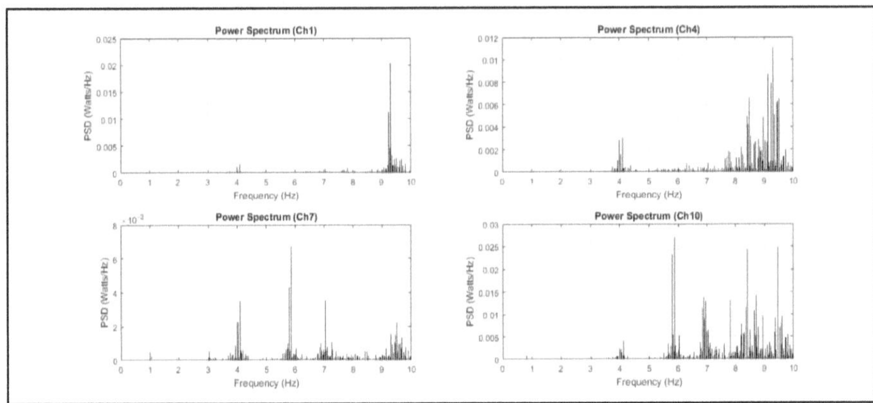

Figure 5-25: Power spectral density of vertical response for Test 2

5.3.5   Comparison between Test 1 and Test 2

Response data recorded during Test 1 showed distinct vibration activity from the footbridge. **Table 5-8** and **Figure 5-26** shows a comparison of the recorded maximum lateral accelerations for the first and last 600s of Test 1. The orange bars in **Figure 5-26** refer to the data recorded in the first 600s and the blue bars refer to the data recorded during the last 600s of the test.

Figure 5-26 shows that the maximum recorded lateral acceleration by each accelerometer was higher in the last 600s than in the preceding 600s. The largest calculated magnitude difference between the two sections of data is observed at accelerometer 8 (i.e. 0.64m/s^2) and the least calculated magnitude difference is observed at accelerometer 5 (i.e. 0.03m/s^2). The overall peak lateral acceleration was recorded by accelerometer 2 (i.e. 1.51m/s^2) during the last 600s of the test while the least peak acceleration was recorded by accelerometer 8 (i.e. 0.74m/s^2) during the first 600s of the same test.

Table 5-8: Maximum acceleration response during first and last 10min of Test 1

Comparison between the first and last 600s of Test 1		
Lateral response		
Ch	First 600s max acceleration (m/s^2)	Last 600s max acceleration (m/s^2)
2	1.0182	1.5070
5	0.8922	0.9247
8	0.7444	1.3892
11	1.1003	1.3549

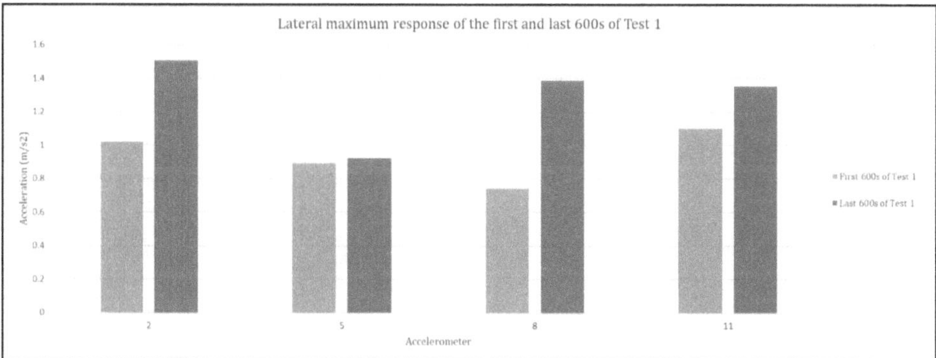

Figure 5-26: Comparison of recorded maximum accelerations for the first and last 600s of the Test 1

Considering what has officially been regarded as Test 1 (i.e first 600s of Test 1), **Table 5-9** compares the maximum recorded lateral accelerations per accelerometer for Test 1 (first 600s) and Test 2. **Figure 5-27** illustrates graphically the maximum accelerations recorded in **Table 5-9**. It is observed that maximum accelerations recorded during Test 1 by accelerometers 2,5 and 11 exceed those recorded during Test 2. The opposite is true for accelerometer 8. The largest magnitude difference is observed at accelerometer 8 with a magnitude of 0.61 m/s^2. The least magnitude difference is observed at accelerometer 11 with a magnitude of 0.029 m/s^2. The overall maximum value recorded is observed at accelerometer 8 for Test 2 with a magnitude of 1.35m/s^2, while the least recorded peak response is observed at accelerometer 8 for Test 1 with a magnitude of 0.744 m/s^2

Table 5-9: Comparison of maximum lateral response between Test 1 and Test 2

Comparison of lateral maximum response of Test 1 and Test 2		
Lateral response		
Ch	Test 1 (first 600s) max acceleration (m/s^2)	Test 2 max acceleration (m/s^2)
2	1.0182	0.7451
5	0.8922	0.7679
8	0.7444	1.3501
11	1.1003	1.0716

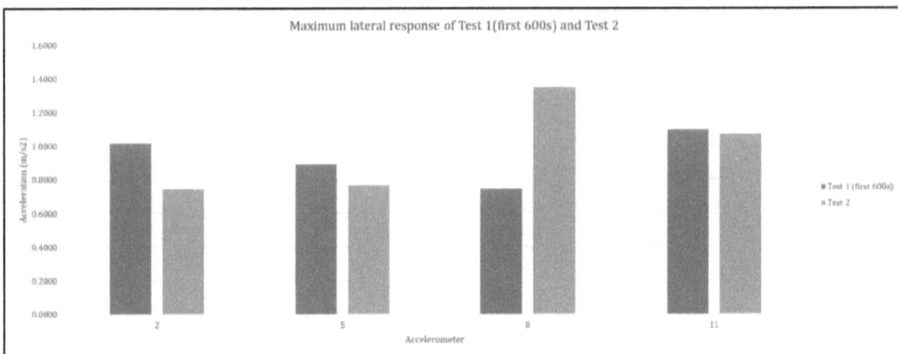

Figure 5-27: Comparison of maximum lateral response between first 600s of Test 1 and Test 2

Chapter 6

6   Discussion

The discussion section seeks to contextualize the results reviewed in the above section by highlighting important observations from the results and contextualizing their relevance to the topic of this book.

6.1   Dynamic characteristics of the Boomslang

The dynamic characteristics of the Boomslang bridge were estimated using a finite element model as well as modal analysis of vibration data collected during a full scale ambient vibration test. It is important to note that the fine tuning of the FE model was beyond the scope of this book and thus obtaining accurate estimations of the natural frequencies and mode shapes was not a priority. Rather, a reasonable estimation of these modal parameters was pursued.

The finite element model showed six natural frequencies between 0Hz and 5Hz (**Figure 5-2**). This was achieved with an assumed target damping ratio of 0.4% which is advised by the Setra and HiVoSS guidelines for steel structures. Timber structures are advised to be modelled with a damping ratio of 3% (Setra, 2006). Since the Boomslang is a composite timber-steel structure with the dominating material being steel, a conservative 0.4% damping ratio was chosen in the analysis. The damping ratio associated with the first mode of vibration was 0.08%. On the other hand, modal analysis results showed ten modes in the range of 0Hz to 5Hz. These natural frequencies were associated with damping ratios between 0.0227% and 0.54%. The fundamental lateral mode was associated with a damping ratio of 0.101%. The additional modes and discrepancies observed in the FE model results compared to that of the operational modal analysis could be attributed to assumptions made concerning the boundary conditions, material properties and geometry of the structural elements (Zivanovi, Pavic & Reynolds, 2006).

Both techniques detected a dominant mode between 0.8Hz and 0.9Hz. Furthermore, closely spaced modes were detected in the range of 0Hz to 5Hz. Evidence of closely spaced modes within this range suggests a high probability of vibration problems instigated by the resonance phenomenon (Setra,

2006) when considering that five harmonics of the pedestrian lateral force occur in this range (Ingólfsson et al., 2011). Evidence of low damping ratios in this frequency range is also an indicator of potential resonance issues (Ingólfsson, 2011).

The first pure lateral modes from the FE model and the ambient vibration results were 0.898Hz and 0.868Hz. This result is not an outlier when compared to other structures of similar span lengths, as illustrated in **Figure 6-1**. This mode is important because it occurs in the detrimental range regarding resonance issues as established by the Setra and HiVoSS guidelines. Moreover, it is important since it appears in the range 0.5Hz-1Hz, which is relevant for the determination of the critical number of pedestrians using the Arup model (Dallard, Flint, et al., 2001). For this mode to be engaged, the walking frequency of the pedestrian would have to be between 1.6Hz and 1.8Hz, representing slow to normal walking pace on the structure. This walking pace is highly probable on the Boomslang given the location and context of the bridge.

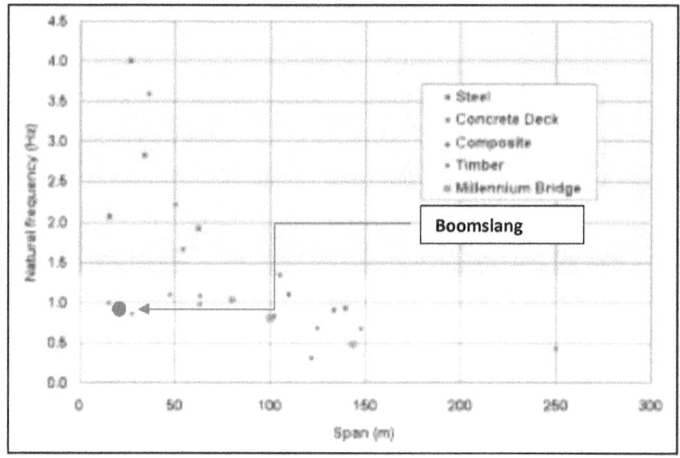

Figure 6-1: Natural frequencies for lateral modes of footbridges (Dallard, Flint, et al., 2001)

6.2   Frequency content of the structure

The importance of determining the operational modal properties of a footbridge, particularly the natural frequencies, is that they indicate whether the structure will be prone to vibration resonance issues or not. These results, along with damping ratio results, are used in developing vibration mitigation devices (i.e damping systems) to maintain the serviceability of the structure under pedestrian loading.

The psd plot results and synchrosqueezed wavelet plot results show the frequency content of the vibration data obtained on the bridge. However, while the psd results show the frequency content in a static form, the synchrosqueezed wavelet plot presents the frequency content in a dynamic form.

The psd plots of Test 1 showed components of 0.86Hz with a dominant intensity of 0.18 at channel 2. A 1Hz component is also observed to be dominant with an intensity of 0.07 at channel 8. On the other hand, the results for Test 2 show a dominant component of 0.80Hz with an intensity of 0.1 at channel 11, while also showing a dominant component of 1Hz with an intensity of 0.13 at channel 8. Comparing the results across the two tests, it is seen that the 0.86Hz component was more engaged during the walk by the groups of students with 10s intervals. On the other hand, the 1Hz component was more engaged during the continuous stream of students test. This suggests that the different cluster forms of the tests influenced the gait, speed and frequency at which the students traverse over the bridge. What is interesting to observe in these results is that the continuous stream of students walked at a higher frequency than the groups of students, which contradicts the notion that the more crowded the walking area is, the lower the frequency of walking will be. However, the result remains reasonable due to the fact that the deck was not completely occupied to the point where normal walking would be difficult to achieve.

The synchrosqueezed wavelet plot for Test 1 presented a reasonably constant wavelet ridge of the 0.86Hz component. The wavelet ridge showed varying intensities at varying sections which were consistent with the moments when the students were walking on or off the bridge. Channel 8, more evidently, shows evidence of a 1Hz component at discrete sections in the time domain. The overall intensity of the wavelet ridge from all channels seems to be approximately 0.8. On the other hand, the synchrosqueezed wavelet plots of Test 2 show two dominant frequencies of 0.80Hz and 1Hz.

These frequencies also appeared with varying intensities at varying sections along the time domain which was consistent with the moments when the students walked on and off the bridge. The general intensity of the wavelet ridge observed in these plots was approximately 0.8.

Contrary to the synchrosqueezed wavelet ridges observed for Test 1, Test 2 plots show distinctly the components of 0.80Hz and 1Hz. From these, it is observed that the 1Hz component is engaged ealier, at approximately 60s, than the 0.80Hz component. The peak intensity of the wavelet ridge for the 1Hz component is observed at approximately 70s while the peak intensity of the 0.80Hz component is observed at approximately 120s. Furthermore, the wavelet ridge following the 1Hz component for Test 2 results is observed to experience a rapid "dip and rise" at approximately 70s. This is at the same instance when the vibration data in the time domain exhibits the intial exponential increase. The phenomenon occurring at this instance cannot be explained yet due to limited data and information but is worth futher investigation.

Comparing the peak intensities reached between the tests, it is reasonable to suggest that the groups of students engaged the first lateral mode of the bridge with more energy than the continuous stream of pedestrians. However, the results from the continuous stream of pedestrians investigation show that multiple frequency components are active during the test. Also, it is distinctly shown that the peak intensity of the engaged frequency is not constant (i.e. the colour is changing) or achieved immediately when the students begin to walk on the bridge. This is also evidence that suggests that SLE and lock-in might not be the initiating mechanisms to excessive lateral vibrations. Rather, there exists another mechanism, prior to SLE and lock-in, that causes the build up of excessive lateral vibrations. From these results, the alluded mechanism itself, and its scientific premise can not be elucidated or proven.

6.3 Pedestrian comfort

Pedestrian vibration comfort on footbridges has become a major concern for footbridge designers. Consequently, increasing research efforts have been directed towards obtaining vibration data on various types of footbridges in order to evaluate their vibration levels and further comment on their serviceability according to the Setra and HiVoSS guidelines. Moreover, vibration comfort is influenced by factors such as the number of people on the bridge, the frequency at which the pedestrians traverse over the bridge, the duration of exposure to the vibrations of the bridge and the severity of the bridge oscillations (Butz et al., 2009). This section will discuss the vibration results obtained from crowd studies conducted on the Boomslang and the implications thereof.

The time domain response results pertaining to Test 1 show two distinct vibration growth and vibration decay sections. All laterally oriented accelerometers show evidence of recorded maximum vibration levels exceeding the minimum comfort limit established by the Setra and HiVoSS guidelines. The highest exceedance to the minimum comfort limit was $0.71m/s^2$ corresponding to 89% exceedance level. According to **Figure 5-8**, the dominant frequency component obtained from the recorded acceleration data is 0.86Hz. This corresponds to a slow to normal walking pace of 1.7Hz which is anticipated on the Boomslang. The guidelines also suggest that 0.86Hz lies within the range of maximum risk of resonance (Setra, 2006), hence the vibration response of the bridge is observed to increase exponentially after some time. If this pace is maintained throughout the walk over the bridge, it is possible that synchronization or lock-in would soon occur and perpetuate the increasing vibration levels of the bridge. Observations of the students walking over the bridge revealed negligible degrees of synchronisation among the students or lock-in of the students and the bridge. However, the vibration response continued to increase exponentially until the students exited the bridge at the opposite end.

The time domain response results of Test 2 show two distinct sections of vibration level increase and decay. Accelerometer 2 and 5 peaked below the minimum comfort limit while accelerometer 8 and 11 peaked beyond the minimum comfort limit. Accelerometer 8 exceeded the minimum comfort limit by $0.55m/s^2$ corresponding to an exceedance level of 69% while accelerometer 11 exceeded the minimum comfort limit by $0.27m/s^2$ corresponding to an exceedance level of 34%. The dominating frequency component in Test 2 was 1Hz. This corresponds to a normal walking pace of 2Hz which is quicker than what is anticipated on the Boomslang bridge. This frequency component lies within the range of maximum risk to resonance according to the design guidelines, hence an increasing level of vibration on the bridge can be observed on the time domain plots (**Figure 5-13**). During the test, negligible degrees of synchronization between the students or the students and the bridge were observed. Again, if this level of vibration is maintained for a longer period, it is highly probable that synchronisation or lock-in might occur on the bridge.

Comparing the recorded peak accelerations of Test 1 and Test 2, it is observed that Test 1 results are consistently higher than Test 2 results, excluding those recorded by accelerometer 8. The highest exceedance level to the minimum comfort limit for Test 1 was 89% compared to that recorded during Test 2 which was 69%. This suggests that the vibration response developed by groups of students with 10s intervals between each group was more severe than that developed by a continuous stream of students. The reason for this could be that the gross area per moment (i.e in time) occupied by the groups of students is larger than that occupied by the continuous stream of pedestrians.

An important observation made in the response results of Test 2 is that during 150s-350s, corresponding to the second walk-over the bridge, a lower peak acceleration was recorded compared to the peak acceleration recorded during 0s-150s. Considering the methodology applied while conducting this test and contrary to the anticipated amplification of the vibration response of the bridge, it is suggested that the reason for the low vibration response during the second walk-over is that the second walk-over by the students was out-of-phase with the vibration mode of the bridge. This suggests that instead of amplifying the vibration response of the bridge (which was the intention of the second walk-over) by inducing the pedestrian dynamic load in-phase with the mode of vibration, the out-of-phase motion of the students dampened the vibration response of the structure.

The Setra and HiVOSS guidelines suggest that $0.1m/s^2$ - $0.15m/s^2$ is the range in which lateral synchronization of pedestrians and lock-in with the bridge vibration behaviour could be initiated. Test 1 results show a gradual increase in vibration response which exceed the $0.15m/s^2$ limit and shows evidence of potential synchronization and lock-in between the students and the bridge. This is noticed at the initiation of exponential growth in vibration response. Similarly, the same vibration response growth trend is noticed in the results obtained from Test 2. The evidence of exponential growth in vibration response is observed beyond $0.15m/s^2$. However, the exponential growth trend distinctly occurs beyond the $0.3m/s^2$ for both tests. Caetano et al., (2010) conducted a pedestrian walk-over test with 145 pedestrians. The vibration response showed evidence of lock-in at $0.2m/s^2$ which confirms the suggestion made by the guidelines of the limit acceleration for a potential lock-in situation. However, $0.2m/s^2$ lies within the mean comfort limit range while $0.3m/s^2$ lies within the minimum comfort limit range. This discrepancy in the initiation of synchronization vibration magnitude supports the notion that the phenomenon of synchronization and lock-in is closely linked to the pedestrians experience of the perceived vibrations (Živanović & Pavić, 2009), rather than the absolute value of the vibrations.

How soon the evidence of exponential growth in vibration response appears is an important factor to consider when attempting to evaluate which of the two test scenarios (representing 2 distinct loading cases) exhibits more concerning results. The results obtained from Test 1 show this evidence between the range of 150s and 200s while the results obtained from Test 2 show this evidence between 50s and 100s. Since the phenomenon of synchronization and lock-in is not strictly a matter of the intensity of the vibrations on the bridge but a matter of the interaction of the pedestrians and the structure, how soon this phenomenon occurs thus becomes more pertinent and primary than the intensity of the vibrations experienced during the phenomenon. Since both test scenarios showed evidence of potential synchronization, then the detrimental case would be that which exhibits this evidence much earlier. Therefore, on the Boomslang, the continuous stream of students is more detrimental than the groups of students.

6.4   Critical number of pedestrians

The concept of the critical number of pedestrians required to trigger synchronization and lock-in on a laterally vibrating pedestrian bridge deck was coined by Dallard et al., (2001) and is formally known as the Arup model. The governing assumption is that there exists a critical number of pedestrians who induce sufficient negative damping force into the bridge system to eliminate the inherent positive damping of the structure. Considering that at resonance conditions, the inherent damping force of the structure is the dominant force maintaining the serviceability of the bridge, eliminating this force could potentially cause the bridge to experience excessive vibrations. Thus, determining this number has recently become an integral part of full-scale footbridge vibration investigations (Ingólfsson & Georgakis, 2011).

The Arup model was used to determine a numerical estimate of the critical number of pedestrians required to trigger lock-in using the numerically determined dynamic modal parameters of the bridge. The estimate was 13 students. Alternatively, the critical number of pedestrians was also determined using the comfort limit criterion established by the reviewed design guidelines. Limited by the influence of unforeseen circumstances during the pedestrian interaction tests, the critical number of pedestrians was not determined in the conventional manner as done by Dallard et al (2001) and Caetano et al., (2010). Rather, the estimated result was a continuous stream of 39 students walking on the bridge for approximately 40s. As seen in **Figure 5-19**, this event is capable of causing instability to the bridge. Furthermore, the response observed in **Figure 5-19** shows that more pedestrians or longer periods of walking could exacerbate the response of the bridge (Brownjohn, Fok, Roche & Omenzetter, 2004).

The popular hypothesis noted in academic literature is that excessive lateral vibrations on a pedestrian bridge are initiated by the SLE mechanism. This notion is challenged by the observations made in this book. The response data clearly shows an exponential increase during certain periods of the test, however, as observed in a short video clip captured by the author, no evidence of pedestrian lateral synchronization or lock-in was observed throughout the recording period. Although SLE is undeniable as a mechanism which perpetuates excessive vibrations on a pedestrian bridge, the author agrees more with the suggestion made by Ingolfsson (2011), Macdonald (2008) and Brownjohn (2004) that there exists another mechanism, prior to SLE, which initiates exponential growth in vibration response which is then perpetuated and sustained by SLE. As suggested by Ingolfsson (2011) and Brownjohn (2004), the initiating mechanism is significantly dependent on the dynamic nature of the damping force throughout the phenomenon. Therefore, more research should be devoted towards understanding the dynamic nature of the damping force, its effect on the stability of the structure and its effect on the critical number of pedestrians.

Given that the result determined in this book is not conventional, the author proposes that it be considered as a lower bound critical number of pedestrians required to trigger lock-in on the Boomslang. The reason for this proposal stems from the observation that 39 students were capable of causing excessive lateral vibrations on the bridge, but the imperative factor was the length of time spent on the bridge which is dependant on the span length of the bridge. By calculation, it was determined that the first student walking at $1.1m/s^2$ for 70s would have walked a distance of 77m on the bridge before the exponential growth in vibration magnitude occured. That is approximately 60% of the bridge span. Therefore, it is vital to evauate whether the last student will be able to exit the bridge before synchronization and lock-in are engaged. Thus, the full span length of the bridge could be considered as an indicator of whether the bridge could possibly experience vibration serviceability issues given a particular density of pedestrians.

Commonly, the critical number of pedestrians is in the range of 100-200 pedestrians (Dallard et al., 2001; Brownjohn, Fok, Roche & Moyo, 2004), however, the probability of that crowd size regularly occupying a bridge at the same time is low. It is highly dependent on major events occurring in the vicinity of the bridge. A crowd size of 39 pedestrians simultaneously walking on a bridge is more probable in a popular area such as Kirstenbosch Gardens.

6.5 Summary of the discussion

The numerical results of the Boomslang bridge show a fundamental lateral frequency of 0.898Hz. The modal analysis results show a 0.868Hz component for the first lateral mode of vibration. These frequencies occur in the maximum risk of resonance range according to the Setra and HiVoSS guidelines. From these results, it is reasonable to conclude that the Boomslang has the potential to experience vibration serviceability issues, including lateral synchronisation and lock-in effects.

Crowd investigations were conducted on the bridge with two scenarios, i.e. groups of students with 10s intervals and a continuous stream of students. The acceleration results show evidence of exponential growth in the lateral direction. The peak acceleration values recorded for both tests exceeded the maximum vibration comfort limit established by the Setra and HiVoSS guidelines. 3 of 4 accelerometers recorded peak acceleration values beyond the minimum comfort limit during Test 1 and 2 of 4 accelerometers recorded peaks accelerations beyond the minimum comfort limit for Test 2. Overall, the results showed that the groups of pedestrians achieved higher peaks of vibration than the continuous stream of pedestrians. Therefore, the groups of pedestrians are a more detrimental loading case on the Boomslang when considering the intensity of the vibrations achieved during this loading case.

The critical number of pedestrians on the Boomslang was evaluated using the Arup model and the Setra comfort limit method. The Arup model evaluated an estimate of 13 pedestrians required to tigger lock-in on the Boomslang. On the other hand, the Setra method evaluated an estimate of 39 students walking on the bridge for 40s. Although the Arup model obtained a conservative result, it is noted that the numerical model was subject to assumptions regarding the stiffness, geometry and materials properties of the structure, thus influencing the modal parameters used as inputs in the Arup model and the estimated result.

The synchrosqueezed plots of the test scenarios conducted on the bridge showed varying degrees of intensity in the time domain. The peak intensity reached in both pedestrian investigations was 0.08. It is noted that the wavelet ridge for the groups of pedestrians' plot first appears with significant intensity after 120s, whereas the wavelet ridge pertaining to the continuous stream of pedestrians' plot first appears with significant intensity at approximately 70s. Also, it is observed that the intensity of the wavelet ridge (i.e. from yellow to red colour) enhances with time. This suggests that the continuous stream of pedestrians would potentially engage the synchronisation and lock-in event earlier than the groups of pedestrians. Therefore, in the context of synchronisation and lock-in, the detrimental loading case would be the continuous stream of pedestrians.

Ultimately, these results logically suggest that the Boomslang is prone to excessive lateral vibrations due to pedestrian lateral loading and that there exists an unknown mechanism (as suggested by **Figure 6-2**), prior to SLE and lock-in, which causes the initation of exponential growth in the vibration response of the bridge. This occurs prior to the physical evidence of synchronisation amongst the pedestrians or lock-in with the vibration motion of the bridge.

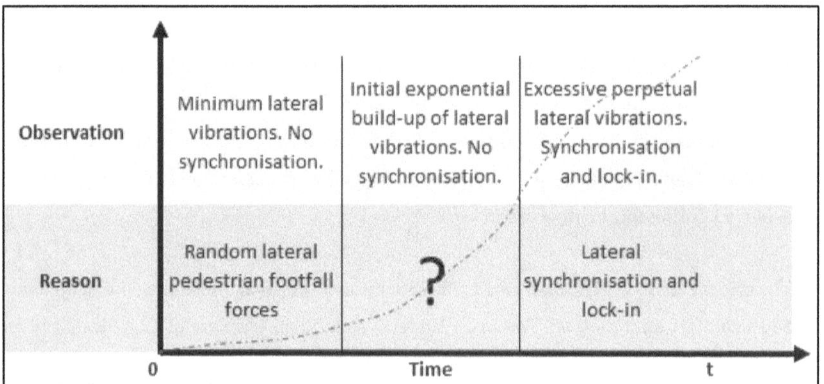

Figure 6-2: A schematic view of the events occuring during the build-up of lateral vibrations on a footbridge. The red dashed line represents the recorded vibration trajectory.

Chapter 7

7   Conclusions

Vibration serviceability of footbridges has become an area of study pursued with much effort due to the structures' primary function, which is to allow pedestrians to traverse across certain obstacles in a comfortable manner(Archbold et al., 2011). The comfort of the pedestrian is influenced by many factors, some of which contribute to the phenomenon of synchronous lateral excitation (SLE). SLE has been widely investigated since it was distinctly observed on the London Millennium Bridge (Dallard, Flint, et al., 2001). The investigations conducted on this bridge lead to the establishment of a formula, formally known as the Arup model, which is capable of determining the critical number of pedestrians required to trigger synchronization and lock-in on a footbridge. Alternatively, widely used guidelines such as the Setra and HiVoSS guidelines have opted to use an acceleration comfort limit to determine the critical number of pedestrians required to trigger synchronization and lock-in on a footbridge. Both these methods of evaluation assume that SLE is the initiating mechanism for excessive lateral vibrations on a footbridge, however, Brownjohn et al., (2004), Macdonald (2008) and Ingólfsson & Georgakis (2011) suggest that there exists another mechanism prior to SLE which initiates excessive lateral vibrations on a footbridge. Currently, this mechanism is only numerical and pending further research to explain the science governing it.

The aim of this book was to investigate the dynamic behaviour of the Boomslang and evalutate the critical number of pedestrians required to trigger synchronisation and lock-in. Furthermore, this book aimed to either support or challenge the notion that SLE is the initiating mechanism for excessive lateral vibrations on footbridges.   The significance of this study is that the Boomslang is a unique, multi-curve canopy walkway situated in a popular botanical garden in Cape Town. Furthermore, the client specifications for the desired footbridge lead the design towards a lightweight and dynamically-active (i.e. lively) solution.

The FE model developed for the Boomslang showed that the fundamental lateral frequency of the bridge is within the detrimental frequency range for resonance as established by the Setra and HiVoSS guidelines. Similarly, the modal analysis conducted on the vibration data obtained on site showed that the operational fundamental lateral frequency of the structure occurs in the detrimental frequency range for resonance as established by the Setra and HiVoSS guidelines.

Two pedestrian loading cases were investigated on the Boomslang. The first test was four groups of students with intervals of 10s between each group and the second test was a continuous stream of students. In both tests, the students were advised to walk the full length of the bridge while adhering to particular gait instructions. The obtained peak acceleration values for both tests showed that the vibration levels experienced on the Boomslang were beyond the minimum comfort limits established by the Setra and HiVoSS guidelines. Furtermore, the results showed that the most detrimental loading case on the Boomslang, when considering the intensity of the vibration levels, is the groups of students. However, the results also showed that the detrimental loading case regarding risk of synchronization or lock-in is the continuous stream of pedestrians.

The critical number of people required to trigger lock-in on the Boomslang was determined using the Arup model and the comfort limit method suggested by the design guidelines. The numerical results suggested that 13 pedestrians are required to trigger synchronisation and lock-in on the Boomslang. Contrary to this, the comfort limit method suggested that 39 pedestrians walking for 40s over the bridge are required to trigger excessive lateral vibrations on the Boomslang. No synchronisation or lock-in was observed during the investigation. This lead to the conclusion that since synchronisation or lock-in were not observed, however, an exponential growth in vibration levels was observed, there should exist another mechanism prior to SLE which causes the initiation of excessive lateral vibrations. This conclusion supports the idea proposed by the Brownjohn et al., (2004), Macdonald (2008) and Ingólfsson & Georgakis (2011).

Lastly, the synchrosqueezed wavelet plots show the engaged frequency contents during the walking of the students in the time domain. The results showed dominant frequency components of 0.80Hz, 0.86Hz and 1Hz. The varying intensity of the wavelet ridge was consistent with the events of the students entering and exiting the bridge. A perculia event was noticed in the Test 2 results which could not be explained yet due to a lack of data and information. Furthermore, an intensity plot bar showed the intensity of the engaged frequency component. The change in colour from blue and yellow to red at particular moments during the walk of the students lead to the conclusion that the degree at which these fundamental modes are engaged increases with time. This was consistent with the observed exponential increase in vibration levels at various moments. This evidence challenges the notion that SLE is the initiating mechanism for excessive lateral vibrations, but rather there is another mechanism, prior to synchronisation or lock-in that causes excessive lateral vibrations.

Recommendations:

The following recommendations are made for further research in this area of study:

- The critical number of pedestrians result obtained using the comfort limit method lead to the hypothesis that there exists a lower bound critical number of pedestrians required to trigger excessive lateral vibrations which could lead to lateral synchronisation and lock-in on a footbridge. It is recommended that this hypothesis be investigated further with the assumption that the lower bound number of pedestrians need to walk half the bridge length at or around the lateral fundamental frequency of the considered span. Its is envisaged that the results obtained from such a study will provide a more practical critical number of pedestrians which could be anticipated as a common traffic volume of the particular footbridge.

- The intiation of excessive lateral vibrations on a footbridge, prior to SLE and lock-in, is attributed to a mechanism yet to be defined. A more insightful explanation of this hypothesised mechanism could be achieved by investigating all three modal parameters found to influence the critical number of pedestrians as per the Arup model (i.e. natural frequency, damping ratio and modal mass) in the time domain. The interaction of these parameters in the time domain could elucidate the intrinsic dynamic behaviour of the footbridge and potentially provide reasons for the development of excessive lateral vibrations.

- The vibration data collected on the Boomslang showed that according to the design criteria established by the guidelines, the Boomslang is a significantly lively footbridge. On the other hand, it is accepted that the footbridge fulfilled the desired client specifications. These contradicting requirements lead to the recommendation that more vibration data should be collected on canopy walkways and additional vibration comfort limits, fitting to the context and functional purpose of these structures, should be established.

Chapter 8

8   Reference list

1. Amezquita-Sanchez, J.P. & Adeli, H. 2016. Signal Processing Techniques for Vibration-Based Health Monitoring of Smart Structures. *Archives of Computational Methods in Engineering.* 23(1):1–15. DOI: 10.1007/s11831-014-9135-7.

2. Archbold, P., Mullarney, B., Mullarney, B. & Archbold, P. 2011. Vibration Serviceability Considerations in Footbridge Design Vibration Serviceability Considerations in Footbridge Design. (June).

3. Bachmann, H. & Ammann, W. 1987. *Vibrations in Structures Induced by Man and Machines.*

4. Brownjohn, J., Zivanovic, S. & Pavic, A. 2008. Crowd dynamic loading on footbridges. In *Footbridge 2008.*

5. Brownjohn, J.M.W., Fok, P., Roche, M. & Moyo, P. 2004a. Long span steel pedestrian bridge at Singapore Changi Airport – part 1: Prediction of vibratio - Articles - The Institution of Structural Engineers. (August). Available: http://www.istructe.org/journal/volumes/volume-82-(published-in-2004)/issues/issue-16/articles/long-span-steel-pedestrian-bridge-at-singapore-cha.

6. Brownjohn, J.M.W., Fok, P., Roche, M. & Omenzetter, P. 2004. Long span steel pedestrian bridge at Singapore Changi Airport – part 2 : Crowd loading tests. *The Structural Engineer.* (August).

7. Brownjohn, J.M.W., Fok, P., Roche, M. & Moyo, P. 2004b. Long span steel pedestrian bridge at Singapore Changi Airport – part 1: Prediction of vibratio - Articles - The Institution of Structural Engineers. *The Structural Engineer.* (August). Available: http://www.istructe.org/journal/volumes/volume-82-(published-in-2004)/issues/issue-16/articles/long-span-steel-pedestrian-bridge-at-singapore-cha.

8. Caetano, E., Cunha, A., Hoorpah, W. & Raoul, J. 2009a. *Footbridge vibration design.*

9. Caetano, E., Cunha, Á., Moutinho, C. & Magalhás, F. 2010. Studies for controlling human-induced vibration of the Pedro e Ines footbridge, Portugal. Part 1: Assessment of dynamic behaviour. *Engineering Structures.* 32(4):1082–1091. DOI: 10.1016/j.engstruct.2009.12.033.

10. Caetano, E., Cunha, Á., Moutinho, C. & Magalhães, F. 2010. Studies for controlling human-induced vibration of the Pedro e Ines footbridge, Portugal. Part 2: Implementation of tuned mass dampers. *Engineering Structures.* 32(4):1082–1091. DOI:

10.1016/j.engstruct.2009.12.033.

11. Caprani, C. 2009. *Structural Dynamics Structural Engineering*.

12. Dallard, B.P., Fitzpatrick, T., Flint, A., Low, A., Smith, R.R., Willford, M. & Roche, M. 2001a. L Ateral V Ibration. *Manager*. 1(December):412–417.

13. Dallard, P., Fitzpatrick, T., Flint, A., Low, A., Smith, R.R., Willford, M. & Roche, M. 2001b. London Millennium Bridge: Pedestrian-induced lateral vibration. *Journal of Bridge Engineering*. 1(December):412–417.

14. Dallard, P., Flint, A., Le Bourva, S., Low, A., Smith, R.M.R. & Willford, M. 2001. The London Millenium Footbridge. *Structural Engineer*. 79(22):17–35. DOI: 10.2749/101686699780481709.

15. Daubechies, I., Lu, J. & Wu, H.-T. 2009. Synchrosqueezed Wavelet Transforms: a Tool for Empirical Mode Decomposition. *Applied and Computational Harmonic Analysis*. 30(2):23. DOI: 10.1016/j.acha.2010.08.002.

16. Feng, Z., Chen, X. & Liang, M. 2015. Iterative generalized synchrosqueezing transform for fault diagnosis of wind turbine planetary gearbox under nonstationary conditions. *Mechanical Systems and Signal Processing*. 52–53:360–375. DOI: 10.1016/j.ymssp.2014.07.009.

17. Fujino, Y. & Siringoringo, D.M. 2015. A Conceptual Review of Pedestrian-Induced Lateral Vibration and Crowd Synchronization Problem on Footbridges. *Journal of Bridge Engineering*. 21(2005):C4015001. DOI: 10.1061/(ASCE)BE.1943-5592.0000822.

18. Fujino, Y., Pacheco, B., Nakamura, S. & Warnitchai, P. 1993a. Synchronization of human walking observed during lateral vibration of a congested pedestrian bridge. *Earthquake Engineering Structural Dynamics*. 22:741–758.

19. Fujino, Y., Pacheco, B.M., Nakamura, S.-I. & Warnitchai, P. 1993b. Synchronization of human walking observed during lateral vibration of a congested pedestrian bridge. *Earthquake Engineering Structural Dynamics*. 22(9):741–758. DOI: 10.1002/eqe.4290220902.

20. Gabor, D. 1946. Theory of communication. Part 1: The analysis of information. *Journal of the Institution of Electrical Engineering*. 93(26):429–441.

21. Gao, R.X. & Yan, R. 2011. From fourier transform to wavelet transform: A historic perspective. In *Wavelets: Theory and Applications for Manufacturing*. 1–224. DOI: 10.1007/978-1-4419-1545-0.

22. Gheitasi, A., Ozbulut, O.E., Usmani, S., Alipour, M. & Harris, D.K. 2016. Experimental and analytical vibration serviceability assessment of an in-service footbridge. *Case Studies in Nondestructive Testing and Evaluation*. 6:79–88. DOI: 10.1016/j.csndt.2016.11.001.

23. Heinemeyer, C. & Feldmann, M. 2008. European design guide for footbridge vibration. *Third International Conference Footbridge 2008: Footbridges for Urban Renewal*. Available: ftp://ftp.fe.up.pt/pub/Pessoal/Dec/manuel/CD Workshop_Pedro/IDW01_Heinmeyer.pdf.

24. Heinemeyer, C., Butz, C., Geradin, M. & Sedlacek, G. 2009. *Design of lightweight footbridges for human induced vibrations*. V. JRC 53442. DOI: 10.2788/33846.

25. Hivoss. 2007. *Human induced Vibrations of Steel Structures Design of Footbridges*.

26. Ingólfsson, E.T. 2011. Pedestrian-induced lateral vibrations of footbridges. Experimental studies and probabilistic modelling. 231(January):281.

27. Ingólfsson, E.T. & Georgakis, C.T. 2011. A stochastic load model for pedestrian-induced lateral forces on footbridges. *Engineering Structures*. 33(12):3454–3470. DOI: 10.1016/j.engstruct.2011.07.009.

28. Ingólfsson, E.T., Georgakis, C.T., Ricciardelli, F. & Jönsson, J. 2011. Experimental identification of pedestrian-induced lateral forces on footbridges. *Journal of Sound and Vibration*. 330(6):1265–1284. DOI: 10.1016/j.jsv.2010.09.034.

29. Ingólfsson, E.T., Georgakis, C.T. & Jönsson, J. 2012. Pedestrian-induced lateral vibrations of footbridges: A literature review. *Engineering Structures*. 45:21–52. DOI: 10.1016/j.engstruct.2012.05.038.

30. Li, C. & Liang, M. 2012. A generalized synchrosqueezing transform for enhancing signal time-frequency representation. *Signal Processing*. 92(9):2264–2274. DOI: 10.1016/j.sigpro.2012.02.019.

31. Macdonald, J.H.G. 2008. Pedestrian-induced vibrations of the Clifton Suspension Bridge, UK. *Proceedings of the Institution of Civil Engineers - Bridge Engineering*. 161(2):69–77. DOI: 10.1680/bren.2008.161.2.69.

32. Magalhães, F., Cunha, Á., Caetano, E. & Brincker, R. 2010. Damping estimation using free decays and ambient vibration tests. *Mechanical Systems and Signal Processing*. 24(5):1274–1290. DOI: 10.1016/j.ymssp.2009.02.011.

33. Mastumoto, Y., Nishioka, T., Shiojiri, H. & Mastuzaki, K. 1978. Dynamic design of footbridges. In *IABSE Proceedings*. 1–15.

34. Matsumoto, Y., Maeda, S., Iwane, Y. & Iwata, Y. 2010. Factors affecting perception thresholds of vertical whole-body vibration in recumbent subjects: Gender and age of subjects, and vibration duration. *Journal of Sound and Vibration*. 330(8):1810–1828. DOI: 10.1016/j.jsv.2010.10.038.

35. Nakamura, S. 2004. Model for lateral excitation of footbridges by synchronous walking.

Journal of Structural Engineering, ASCE. 130(1):32–37. DOI: doi:10.1061/(ASCE)0733-9445(2004)130:1(32).

36. Nakamura, S. ichi & Kawasaki, T. 2006. Lateral vibration of footbridges by synchronous walking. *Journal of Constructional Steel Research*. 62(11):1148–1160. DOI: 10.1016/j.jcsr.2006.06.023.

37. Newland, D.E. 2004a. Pedestrian excitation of bridges. 218(July 2003):477–492.

38. Newland, D.E. 2004b. Pedestrian excitation of bridges. *Journal of Mechanical Engineering Science*. 218(5):477–492. DOI: doi:10.1243/095440604323052274.

39. Newland, D.E. 2004c. Pedestrian excitation of bridges. *Proceedings of the Institution of Mechanical Engineers Part C - Journal of Mechanical Engineering Science*. 218(5):477–492. DOI: doi:10.1243/095440604323052274.

40. Newland, D.E. n.d. Pedestrian Excitation of Bridges - Recent Results. (1):1–15.

41. Van Nimmen, K., Lombaert, G., De Roeck, G. & Van den Broeck, P. 2014. Vibration serviceability of footbridges: Evaluation of the current codes of practice. *Engineering Structures*. 59:448–461. DOI: 10.1016/j.engstruct.2013.11.006.

42. Pedersen, L. & Frier, C. 2010. Sensitivity of footbridge vibrations to stochastic walking parameters. *Journal of Sound and Vibration*. 329(13):2683–2701. DOI: 10.1016/j.jsv.2009.12.022.

43. Peeters, B. 2000. System identification and damage detection in Civil Engineering.

44. Perez-ramirez, C.A., Amezquita-sanchez, J.P., Adeli, H., Valtierra-rodriguez, M., Camarena-martinez, D. & Romero-troncoso, R.J. 2016. New methodology for modal parameters identification of smart civil structures using ambient vibrations and synchrosqueezed wavelet transform. *Engineering Applications of Artificial Intelligence*. 48:1–12. DOI: 10.1016/j.engappai.2015.10.005.

45. Rainieri, C. & Fabbrocino, G. 2014. *Operational Modal Analysis of Civil Engineering Structures*. DOI: 10.1007/978-1-4939-0767-0.

46. Ren, W.X. & Peng, X.L. 2005. Baseline finite element modeling of a large span cable-stayed bridge through field ambient vibration tests. *Computers and Structures*. 83(8–9):536–550. DOI: 10.1016/j.compstruc.2004.11.013.

47. Ren, W., Zhao, T., Harik, I.E. & Asce, M. 2004. Experimental and Analytical Modal Analysis of Steel Arch Bridge. 130(July):1022–1031.

48. Rent, W.X. & Zong, Z.H. 2004. Output-only modal parameter identification of civil engineering structures. *Structural Engineering and Mechanics*. 17(3–4):429–444. DOI:

10.12989/sem.2004.17.3_4.429.

49. Reynolds, P. 2014. Model testing of a 34m catenary footbridge. (January 2002).

50. Ricciardelli, F. & Pizzimenti, A.D. 2007. Lateral Walking-Induced Forces on Footbridges. *Journal of Bridge Engineering*. 12(December):677–688. DOI: 10.1061/(ASCE)1084-0702(2007)12:6(677).

51. Roberts, T.M. 2005. Lateral Pedestrian Excitation of Footbridges. 10(February):107–112.

52. SAISC Projects. 2015. The Boomslang. *Steel Construction*. 39(3):24–25.

53. Setra. 2006. *Technical guide Footbridges Assessment of vibrational behaviour of footbridges under pedestrian loading.*

54. Shahabpoor, E., Pavic, A. & Racic, V. 2017. Structural vibration serviceability: New design framework featuring human-structure interaction. *Engineering Structures*. 136:295–311. DOI: 10.1016/j.engstruct.2017.01.030.

55. Venuti, F. & Bruno, L. 2009. Crowd-structure interaction in lively footbridges under synchronous lateral excitation: A literature review. *Physics of Life Reviews*. 6(3):176–206. DOI: 10.1016/j.plrev.2009.07.001.

56. Wang, P., Gao, J. & Wang, Z. 2014. Time-frequency analysis of seismic data using synchrosqueezing transform. *IEEE Geoscience and Remote Sensing Letters*. 11(12):2042–2044. DOI: 10.1109/LGRS.2014.2317578.

57. Yoshida, J., Fujino, Y. & Sugiyama, T. 2007. Image processing for capturing motions of crowd and its application to pedestrian-induced lateral vibration of a footbridge. *Shock and Vibration*. 14:251–260. DOI: 10.5100/jje.42.279.

58. Zivanovi, S., Pavic, A. & Reynolds, P. 2006. Modal testing and FE model tuning of a lively footbridge structure. 28:857–868. DOI: 10.1016/j.engstruct.2005.10.012.

59. Zivanovi, S., Pavic, A. & Ingolfsson, E.T. 2010. Modeling Spatially Unrestricted Pedestrian Traffic on Footbridges. 136(October):1296–1308.

60. Živanovi, S., Pavic, A., Reynolds, P. & Vujovi, P. 2005. Dynamic analysis of lively footbridge under everyday pedestrian traffic. In *EuroDyn*. 453–459.

61. Zivanovic, S., Pavic, A. & Reynolds, P. 2005. Vibration serviceability of footbridges under human-induced excitation: A literature review. *Journal of Sound and Vibration*. 279(1–2):1–74. DOI: 10.1016/j.jsv.2004.01.019.

62. Živanović, S. & Pavić, A. 2009. Probabilistic assessment of human response to footbridge vibration. 28(4):255–268. DOI: 10.1260/0263-0923.28.4.255.